Donald Wortzman

The Gyroverse

The Hidden Structure of the Universe

Donald Wortzman

The Gyroverse

The Hidden Structure of the Universe

By

Donald Wortzman

Donald Wortzman

The Gyroverse
The Hidden Structure of the Universe

© 1999 - 2023 by Donald Wortzman.
All rights reserved.

No part of this book may be reproduced, stored in a retrieval system, or transmitted by any means, electronic, mechanical, photocopying, recording, or otherwise, without written permission from the author.

ISBN-13: 9781450516174

Expanded Fourth Edition

http://www.amazon.com/dp/1450516173/

CreativeTomes – Post 178 – 11/28/2022

Donald Wortzman

The Gyroverse
The Hidden Structure of the Universe

To My Grandchildren

Donald Wortzman

PREFACE

This book is a work-in-progress dealing with an all-encompassing physics theory, sometimes referred to as a "Theory of Everything." It began in 1999 as part of a larger work that was to contain material on various subjects. Having just retired, I spent a couple of hours each afternoon in bookstore cafes. These seem to be reincarnations of the old European tradition of the coffee house being a gathering place for discussion. Barnes and Noble stores encourage this connection with their mural of the great writers, Faulkner, Steinbeck, Eliot, and others, in their café. Mornings, I would post progress from my notes of the previous day on my Web site, gyroverse.com. The idea was exhilarating that for only a nominal cost, anyone around the world could access my Website. After a while, I realized that most of my updates were mainly concerned with physics, and decided to concentrate on it, postponing the rest for a later date. Since the site's inception, it has received about 20 thousand hits. Much of this traffic originated from two Web sites that erstwhile featured new theories outside the current paradigm, Physlink and Science Nook.

Though my formal training has been in mathematics and engineering, physics now became my passion and new post-retirement vocation. I read voraciously on the subject, but became increasingly frustrated by the lack of descriptions of how anything actually functioned. Could it be that the physics community was stymied; did not believe there was an underlying physical cause; stopped looking; or did not to care? The Nobelist Richard Feynman said it succinctly, "...the more you see how strangely nature behaves, the harder it is to make a model that explains how even the simplest phenomena actually work. So theoretical physics has given up on that." Current physics can be thought of as God's magic trick, where physicists are spectators satisfied not knowing how the trick is done.

Feynman's observation seemed outrageous to me, I believed someone could correct it, and wanted to give it a try. While it might appear to be a monumental undertaking, especially for a layperson to attempt this, there is a tradition of scientific dilemmas being solved by unknowns. Even the incomparable Einstein, the most prolific physicist in modern times, was an obscure Swiss patent clerk when he made his most famous discoveries. Revamping physics was like solving a giant jigsaw puzzle. Many pieces fit together nicely, from context, some could be approximately placed, and others on the sideline still wait to be positioned. After working with the material for 3 years, with the puzzle far along, I decided in March 2002 to publish my results in the first edition of the Gyroverse. My book was made available through all the normal bookseller channels, as well as directly from the publisher, in paperback, and e-copy.

I once thought that physics was the purest of the sciences and engineering was applied physics. However, from this project, I came to realize that the design of the universe was replete with the same engineering principles and practices used here on Earth, designing reliable systems with less reliable parts. It contains examples of feedback loop control, both positive and negative, mechanical advantage, ratio control, energy storage, memory, and motion stabilization. While the physics parameters were created for, and limited to this universe, the design principles seem to apply more broadly. Therefore, in retrospect, it is not odd that an engineer would undertake this study.

My greatest challenge is to have this theory seriously reviewed by the physics establishment. It should not have been a surprise, because even recognized physicists have this difficulty. However, it

The Gyroverse
The Hidden Structure of the Universe

is encouraging that others who received this treatment still prevailed. In the fascinating book <u>Faster than the Speed of Light,</u> cited in the bibliography, physicist João Magueijo, an Imperial College physics professor discussed his Variable Speed of Light Theory, an alternative to inflation theory. It challenged the most basic tenet of relativity, the constancy of the speed of light. In the book, he discussed being marginalized by the physics establishment, and perhaps jeopardizing an academic future by his persistence in trying to have that idea heard. He finally prevailed, not that the theory has been elevated to dominance, but at least it is now considered a legitimate speculative alternative. I especially empathized with him, since the gyroverse theory also challenges the current paradigm. Being a complete outsider makes it even more difficult to be heard, but at this stage of life, fortunately I have the freedom to pursue my quest.

After the book was published, I continued to work on this project, adding pieces to the puzzle, enlarging it by 50 percent in the second edition, and another 15 percent in this third edition. One idea developed is the evolution of the universe, from inception to its current embodiment. The big bang was more like a big push that is still progressing. Another is the mechanism behind particle spin. In addition, many explanations have been sharpened. Several important anomalies have been explained. Finally, the thought-experiment to show that information can travel faster than the speed of light has been modified twice, because of revisions to the earlier ideas. With all these changes, it seemed that it was time to present the current status in an expanded third edition.

At all significant junctures in developing this treatise has been copyrighted. This was done to leave an audit trail for the evolution of the thought process. At each juncture, new material was added, but also some changes were made. As mentioned, this is a work in progress where the end is still in the future. Some remaining rough edges suggest the need for more fine-tuning that will continue. Whenever a refinement is made, it must be consistent with everything that preceded it. At its present incarnation, it can tie most aspects of quantum theory, relativity, and cosmology together,

Donald Wortzman

under one umbrella, and clarified much that has defied explanation. An advantage in this latest edition, on CreateSpace, is that the book can still be revisited, without having to republish, which I have continually done, and still do.

The Gyroverse
The Hidden Structure of the Universe

WHAT'S WRONG WITH PHYSICS

Often it is said that the current physics theories have been through many tests, and each time it re-confirmed the current paradigm. Nothing could be further from the truth. Inconsistencies are ignored, until some outrageous side theory is devised that allegedly explains it, allowing the pretext to continue. The following is a partial list that highlights what's wrong with physics, all of which are addressed and better explained in this treatise.

Inflation Theory: General Relativity, as originally envisioned, could not explain the uniformity of matter in the universe and the uniform cosmic background radiation, a remnant from the big bang. To obviate this dilemma, inflation theory, which is nothing like relativity and lacks any independent evidence, was invented.

Expansion Theory: There is no evidence of the kinetic energy that should be accompanying the expanding universe. In addition distant galaxies moving away from us close to the speed of light should appear packed closer together. Thus, expansion theory was devised that posits that nothing is really moving, but instead empty space is being created between the galaxies.

Illusive Dark Matter: The universe is shy of much too much visible matter to explain why galaxies hold together, and why light bends (lensing) so much around them. Consequently, it was concocted that 85 % of the matter in the universe is dark matter that can't be seen or otherwise detected, not even one atom's worth, yet nevertheless exerts gravitational pull...aether redux.

Accelerated Expansion Surprise: Analysis suggests that the universe is not simply expanding but that the expansion is accelerating. To explain this surprising result an ad hoc constant of integration has been added to the General Relativity equations, representing dark energy. To this day, there has been no direct validation of this dark energy fudge factor. The only theory that allegedly might apply predicts many orders of magnitude more acceleration.

Entangled Particles Illusion: These particles are able to stay in contact miles apart. The mechanism for so doing remains strange and physically unexplainable. Einstein even called it "spooky action at a distance." While it is real, it's still an unresolved physical mystery. The Copenhagen Interpretation institutionalized the notion that there is no physical explanation for physics.

Twin Slits Experiment: Atomic particles passing through dual slits one at a time make individual dots on a photo sensitive screen behind the slits, indicating that the particle has passed through only one of the slits. Nevertheless, the accumulation of these dots form a diffraction pattern indicating that each photon must have passed through both slits simultaneously, interfering with themselves. Both explanations can't be simultaneously true, without the fiction that its the inanimate probability that is doing the interfering.

Allais Effect: The gravitational pull on earth by the sun and moon combination increases as the two approach each other before an eclipse. This is expected, because both are then pulling in the same direction. However, when the eclipse actually occurs their gravitational pull mysteriously dips.

Pioneer 10 & 11 Probes Mystery: These space shots on leaving our solar system are being slowed by the sun's gravitational pull much more than would be expected. This ignored anomaly had no explanation consistent with current theory. Recently, it was blamed on asymmetrical thermal radiation. This was suggested previously, but this time it was done with a sharper pencil. It seems that the establishment is much too eager to accept it.

Mass Value Anomaly: The mass of inertia and gravity, two seemingly unrelated phenomena, surprisingly have the same value. This unresolved mystery dates back over 300 years to Newton, and is still no closer to being explained. More recently, this anomaly could be extended to the mass of energy, which is on the surface very different from the others.

Matter Energy Equivalence: $E=MC^2$, the conversion of matter to energy, is strange because M is also the mass of inertia and E has the units of kinetic energy without any apparent motion. It should be expected to have an unrelated formula.

Electron-Nucleus Preclusion: Why aren't electrons ever pulled into the nucleus of atoms, since both are oppositely charged and are attracted to each other?

Non-Simultaneity Illogicality: Relativity posits that if two observers are in relative motion, A should observe B's clock going slower than its own. However, B should likewise observe A's clock going slower than its own. This reciprocity was invalidated in tests conducted on airplanes and satellites, challenging the notion of non-simultaneity.

Ehrenfest Paradox: There is no evidence that the length of objects shorten when in motion, as relativity predict. In fact, early on, the Ehrenfest showed that length shortening is problematic.

Delayed Choice Conundrum: According to accepted quantum theory interpretation, the delayed choice experiment, originally devised by John Wheeler, could demonstrate time going backwards. While in a lab setup it seemingly went backwards only a small fraction of a second, according to the original proposal, it can theoretically go back billions of years.

Gravitation-Acceleration Equivalence: This assertion, the cornerstone of general relativity, maintains that no experiment can be constructed to distinguish between the two. The fact that light bends twice as much by gravity than by acceleration makes the assertion patently false.

Stellar Aberration: Special relativity dictates that all motion is relative and no experiment can determine which body is actually moving. The same amount of stellar aberration, independent of a stars motion, for one, violates this principle.

Gravitational Lensing: The gravitational bending of light, one of the pillars of the acceptance of general relativity that captured the imagination of the world can be derived using only classical physics arguments.

Mercury's Perihelion Advance: The second cornerstone, possibly more persuasive to the physics community, was the calculation of Mercury's perihelion advance. This was less dramatic because it only involved explaining phenomena that was already known. Nevertheless, this calculation can now be done without invoking general relativity.

Forces of Attraction: There is no physical argument to justify why two separated atomic particles or celestial bodies should be attracted to each other. The physics jargon that the force is mediated by exchanging force particles conveys no credible information on how or why it happens.

Higgs Boson: The illusive Higgs particle, which supposedly would explain the origin of mass, and also gives all the force particles their oomph had been searched for to no avail. In addition, relating the graveon to all the other force particles has also not occurred. The new particle accelerator, Large Hadron Collider (LHC) allegedly found the Higgs, but it was not shown that it actually caused mass.

Speed of Gravity: General relativity posits that the speed of gravity is equal to the speed of light. The lack of gravitational aberration from the sun gives ample evidence to suggest that gravitation's speed is much faster than lightspeed.

Anti-Matter Dilemma: The nature of anti-matter is one of the big mysteries of physics. Why an explosion ensues, annihilating both particles, when matter and anti-matter collide is physically unexplainable.

Particle Spin: Particle spin is another baffling mystery of physics. Since the spin velocity is so great that the outer shell of the particles would be moving much faster than lightspeed, it was dubiously decided that spin is purely a quantum effect with nothing actually spinning. It is said to have no classical counterpart, as if that jargon clears it up.

String Theory: This theory, which has been in forefront for thirty years with thousands of physicists working on it, has very little to show for itself. It is time to cut our losses, and vigorously pursue other possibilities. This theory makes better use of extra dimensions.

Anthropic Principle: The parameters governing the workability of the universe are so precariously balanced that it had to be assumed that this universe is only one of an infinity of universes, all with different sets of parameters. The argument that this universe just happens to have those that allowed life to form has been dubbed the anthropic principle. Now, every time new science (e.g. string theory) can't be adequately explained, it becomes additional corroborative evidence supporting the anthropic principle…if you can't fix it, feature it.

Relativity-Quantum Incongruity: It is widely acknowledged that relativity and quantum theories are at odds. But, actually it's worse, since cosmology theory is not aligned with either. Inflation, expansion, dark matter, and dark energy are peculiar to cosmology, and are not predicted by the other theories, or reproducible in the laboratory.

Lightspeed: Why are both matter and photons coincidentally limited by the same speed of light? Also, what is the mechanics of the limiting factor, and the connection between them?

Sagnac Effect: When pulses of light are transmitted simultaneously around earth both with and opposite Earth's rotation, and they meet at the point of transmission, the one transmitted in the direction of rotation arrives first. Their difference is twice the speed of the earth's surface rotational speed of about 1000 miles per hour, indicating that the speed of light is not always exactly "c." Since the trip completely around earth is different from "c," all shorter segments must also differ from "c." In other words, most light segments on earth, don't travel at the exact speed "c." Relativity enthusiasts argue that the earth is rotating and is not an inertial frame, so inertial frame rules do not apply.

However, a 10 mile segment would vary insignificantly from an inertial frame, so according to special relativity the speed in any direction on this segment should be the same. It is not necessary to know the velocity of this quasi inertial frame relative to anything else according to special relativity. At the very least, it is a big enough circle, so one would expect its relative speed fall somewhere in between an inertial frame and a rotating body, so that its directional difference be something less than twice Earth's rotational speed.

In addition, earth is also rotating around the sun at 67 thousand miles per hour, and circulating with the galaxy at 575 thousand

The Gyroverse
The Hidden Structure of the Universe

miles per hour. A 10 mile segment of these rotations are even much closer to an inertial frame. There is no reason to expect that the Sagnac effect does not apply to them, leaving most light not actually traveling at exactly 'c." In principle, there are no true inertial frames, so that every test confirming Special Relativity would be invalid and just a fortuitous accident. Even general relativity is problematic since it was designed to default to special relativity.

The point has been reached where we must accept that something is very wrong with our notion of physics, where the simplest observables can't be explained. The physics community has become comfortable with this rationalization, but familiarity is not a substitute for understanding. It can't be fixed by just tweaking the current paragon. The gyroverse is a completely new physics paradigm that makes sense from all this drivel. It resolves these issues and much more.

Donald Wortzman

The Gyroverse
The Hidden Structure of the Universe

ACKNOWLEDGMENT

I wish to express my gratitude for the help I received in writing this book. Members of the Mahopac Library Writers' Group were especially helpful. Although physics was far afield of any of their expertise, they persevered and gave invaluable assistance, especially in the non-technical areas. In particular, I would like to mention Vincent Dacquino who led the group and provided much needed encouragement. In addition, I owe a debt of gratitude to my wife Margie for editing the final manuscript, as well as her support over the years in writing the book.

Donald Wortzman

TABLE OF FIGURES

Figure 3.1 Four-Dimensional Hyper-Box
Figure 3.2 Flatlander Soldier Class
Figure 3.3 Different Flatliner's Terrain

Figure 4.1 Representation of One Axis
Figure 4.2 Three-Dimensional Space Cells
Figure 4.3 The LeSage Mechanism
Figure 4.4 The Gyroscope Mechanism
Figure 4.5 Gyro-Vectorial of The Fourth Axis
Figure 4.6 Gravitational Deflection of Light

Figure 5.1 Simultaneity I - Relativity
Figure 5.2 a/b, Simultaneity II - Gyro Vector
Figure 5.3 a/b, Simultaneity III - Bullet
Figure 5.4 Binary Star System
Figure 5.5 a/b Binary Star Vector Diag
Figure 5.6 Velocity Transformation
Figure 5.7 Space Travel Vect Diag

Figure 6.1 Representation of One Axis
Figure 6.2 Three-Dimensional Space
Figure 6.3 Double Slit Experiment
Figure 6.4 The Two Photon Paths

Donald Wortzman

Figure 6.5 Graveon Derivative Forces

Figure 7.1 Three Universe Expansion Scenarios
Figure 7.2 The Universe Model
Figure 7.3 Virtual Location of Objects
Figure 7.4 Virtual & Actual Path of Light
Figure 7.5 Gravitational Lensing by Galaxy
Figure 7.6 Allais Effect
Figure 7.7 Bonding Mass Decrease
Figure 7.8 Cosmic Delayed Choice Experiment
Figure 7.9 Delayed Choice Experiment
Figure 7.10 A Time Slice of The Universe
Figure 7.11 Inside vs Outside Slice of the Universe

Figure 8.1 Michelson-Morley Setup
Figure 8.2 Fizeau Experiment Setup
Figure 8.3 The Proverbial Special Relativity Train

Figure 10.1, Parallax Shift
Figure 10.2, Three Universe Expansion Solutions

Figure 12.1, Instant Messaging Configuration
Figure 12.2, Phase Splitter

TABLE OF CONTENTS

TO MY GRANDCHILDREN IX
PREFACE XI
WHAT'S WRONG WITH PHYSICS XV
ACKNOWLEDGMENT XXIII
TABLE OF FIGURES XXV
TABLE OF CONTENTS XXVII

PART I - THE GYROVERSE MODEL 37

CHAPTER 1 - INTRODUCTION 38
CHAPTER-BY-CHAPTER OVERVIEW 41
Re: Chapter 1 - Introduction 41
Re: Chapter 2 – The Problems 42
Re: Chapter 3 - The Model Basics 42
Re: Chapter 4 - The Gyroverse Model 43
Re: Chapter 5 – Relativity Implications 44
Re: Chapter 6 - Quantum Implications 45
Re: Chapter 7 - Cosmology Implications 46
Re: Chapter 8 - Relativity Overview 48
Re: Chapter 9 - Quantum Overview 50
Re: Chapter 10 - Cosmology Overview 52
Re: Chapter 11 - Triplet Paradox 52

Re: Chapter 12 – Instant Messaging 53
CHAPTER 2 - THE PROBLEM 54
  INTRODUCTION 54
  RELATIVITY ENIGMAS 59
    Energy-Matter Equivalence ($E = mc^2$) 59
    Inertia, Gravity, and Energy Equivalence 61
    Relativistic Non-Simultaneity 62
    Relativistic Distance Contraction 63
    Gravitation-Acceleration Equivalence 63
    Bending of Light Prediction 64
    Gravitational Force Duality 65
      Space-Time Warp 66
      Elusive Graveon Particle 67
      Immediateness of Gravitational Force............. 68
  QUANTUM ENIGMAS 70
    Electromagnetic Force 70
    Particle/Wave Duality I 71
    Particle/Wave Duality II 72
    String Theory 73
    Quantum Entanglement 75
  COSMOLOGY ENIGMAS 76
    Universe Inflationary Model 76
    Universe Expansion Model 78
  SUMMARY 79
    Summarizing Enigmas 79
    Predictions - Postdictions 80
CHAPTER 3 - THE MODEL BASICS 82
  Introduction 82
  Ball of String Metaphor 82
  Card File Example 83
  Hyper-Box Simile 84
  Edwin Abbott's Flatlanders 87
  Gyrohelix Precursor 89
  Cardinality Digression 90
  Introducing the Gyrohelix 90
CHAPTER 4 - THE GYROVERSE 92

Introduction	92
Gyroverse Model	93
GRAVITY MECHANISM	**100**
The Graveon	100
Gravitational Force - LeSage Theory	102
INERTIAL MECHANISM	**105**
The Gyrohelix	105
The Gyroscope Operation	106
Effective Inertia Amplification	108
Creating Inertia	110
Intrinsic Inertia - Graveon Connection	111
The Fourth Space Dimension	113
Relativistic Time and the W-Axis	116
Gyroverse Energy	117
GAMMA FACTOR	**118**
Gamma Relationship to Velocity	118
Gravitational Effect on Light	120
Gravitational Effect on Light Calculation	121
Modified Newtonian Dynamics Revisited	126
Modified Centrifugal Acceleration	129
Other Effects of Increasing Gamma	129
CHAPTER 5 - RELATIVITY IMPLICATIONS	**131**
NON-SIMULTANEITY	**131**
Einstein's Train Thought-experiment	131
Speeding Bullet Substitution	137
Speed of Light – Binary Stars	140
GYROVERSE CHARACTERISTICS	**144**
Yaw, Pitch, and Roll	144
Distance	146
Time, A Byproduct of Inertia	147
Mass and Energy	148
Speed of Light	148
LORENTZ TRANSFORMATION	**149**
Lorentz Transformation Chart - Gyro	151
Coordinates	152

- Time Dilation 152
- Velocity Addition 153
- Velocity Addition Interpretation 155
- Lightspeed Velocity Limitation 156
- Relative Velocity of Inertial Frames 157
- Mass, Momentum, and Energy 157
- Kinetic Energy From Gamma Factor 158
- Kinetic Energy Derivation 158
- Total Energy is mc^2, not $1/2(mc^2)$ 159
- Doppler Effect 160
- GYROVERSE POTPOURRI 160
  - Streamline Effect 160
  - Objects in Position-space 161
  - Space Travel 161
  - Ehrenfest Paradox 164
  - Muon Decay 165
- CHAPTER 6 - QUANTUM IMPLICATIONS 167
  - ENTANGLEMENT AND CLOSENESS 167
  - WAVE-PARTICLE DUALITY 171
  - ELECTROMAGNETIC FORCE 175
    - Photons Dual Paths 175
    - Determining Dual Path 178
  - ANTIMATTER 178
    - Traditional Antimatter Theory 179
    - Gyroverse Antimatter Theory 181
  - GRAVEON FORCE DERIVATIVES 183
    - The Strong Force 183
    - The Weak Force 186
    - Entanglement Force 187
    - Composite Particle Mass 188
    - Fermi & Bose Particle Statistics 188
  - PARTICLE SPIN 190
    - Historical Perspective 190
    - Atoms Physical Structure 191
    - Additional Quantum Numbers 192
    - Electron Spin History 194

Electron Spin	195
The Planck Constant	200
de Broglie Wavelegnth	201
ATOM FORMATION	202
STRING THEORY	202
CHAPTER 7 - COSMOLOGY IMPLICATIONS	204
INTRODUCTION	204
COSMOLOGICAL MODEL	206
The Flatness and Horizon Problems	208
Flatness Problem	208
Horizon Problem	209
Inflation Theory	210
The Gyroverse Big Push Model	211
Introduction	211
Possible Creation Scenario	212
Universe Expansion	213
Photons and Look Back	218
Uniform Expansion	219
Size, Look and Energy	221
Matter Formation	222
Frequency Shift in Star Radiation	223
Age of the Universe	224
Universe Pseudo-Acceleration	225
COSMOLOGY POTPOURRI	228
Dark Matter Hypothesis	228
Modified Newtonian Dynamics	230
Pioneer Mystery and Allais Effect	235
Flyby anomoly	238
Bonding Energy	239
Stellar Aberration	241
Delayed Choice Experiment	242
Dimensions of Matter	248
Hidden Variables	249
CMB Dipole Anisotropy	250
Other Universe Characteristics	251

CONCLUSION 252
   Structural Evolution of the Universe..................254
   Evidence For More Than Three Dimensions 255

PART II - CURRENT PHYSICS OVERVIEW............265

CHAPTER 8 - RELATIVITY OVERVIEW..........................266
INTRODUCTION...266
GALILEAN TRANSFORMATION.......................267
- Galilean Transformation Chart............................268
- Coordinates..269
- Time Translation 269
- Velocity Transformation 269
- Mass, Momentum, and Energy............................270
- Doppler Effect 270
- General 270

LIGHTSPEED EXPERIMENTS 271
- Maxwell's Equation 271
- Aether (Ether)..271
- Michelson-Morley experiment 272
- The Fizeau Experiment..274
- Lorentz Transformation.......................................276

EINSTEIN'S AMAZING YEAR............................277
- Photoelectric Effect 277
- Brownian Motion 278
- Avogadro's Number..278
- Special Relativity 279
  - Postulate I - Principle of Relativity 279
  - Postulate II - Lightspeed Constant 280
  - Additional Requirement 280

LORENTZ TRANSFORMATION - SR.................280
- Lorentz Transformation Chart – S. R. 281
- Coordinates..282
- Time Dilation and Non-Simultaneity 282
- Transitivity Law & Non-Simultaneity 284
- Velocity Transformation 287
- Mass, Momentum, and Energy............................287

- Doppler Shift 288
- Gamma Factor 288
- GENERAL RELATIVITY 289
  - Mach's Principle 289
  - General Relativity Verification 290
    - Advance of Mercury's Perihelion 291
    - Bending of Starlight by the Sun 292
- CHAPTER 9 - QUANTUM THEORY OVERVIEW 293
- EARLY HISTORY 293
  - Introduction 293
  - Light: A Particle or a Wave 294
  - Initial Atom Theory 294
- QUANTUM THEORY 294
  - Origin 294
  - Electron Spin 297
  - Particle - Wave Duality 298
  - Schrödinger's Cat 299
- BELL'S THEOREM 300
  - Entangled Particles 300
  - EPR Paradox 301
  - Nicolas Gisin Experiment 302
- THE STANDARD MODEL 305
  - The Atom 305
    - Nucleus 305
    - Electrons 306
    - Elementary Particles 306
    - Fermions and Bosons 307
  - Four Forces of Nature 307
    - Gravity 307
    - Electromagnetism Forces 308
    - Weak Force 308
    - Strong Force 308
  - Quantum Electrodynamics (QED) 309
  - Quantum Chromodynamics (QCD) 310
  - Electroweak Force 310

Higgs Field and Higgs Boson 310
STRING THEORY 311
  Introduction 311
  Gunnar Nordström Theory 311
  Kaluza-Klein Theory 312
  String Theory and M-Theory 313
  Symmetry and Groups 315
CHAPTER 10 - COSMOLOGY OVERVIEW 317
ANCIENT COSMOLOGY 317
CLASSICAL COSMOLOGY 319
  Nicolas Copernicus 319
  Galileo 320
  Kepler 320
  Isaac Newton 321
STANDARD COSMOLOGY MODEL 322
  General Relativity 322
  Hubble Constant 323
  Hubble Constant Update 324
  Big bang Theory 325
  Stellar Distance 325
    Parallax Shift 326
    Standard Candle-Cepheid Variables 327
    Standard Candle-Supernova Type 1a 328
    Doppler & Expansion Shift 329
  Inflation Model 330
    Flatness Problem 330
    Horizon Problem 331
  Accelerated Universe Expansion 334

PART III – THOUGHT EXPERIMENTS 337

CHAPTER 11 - TRIPLETS PARADOX 338
  Introduction 338
  Scenario 338
  Conclusion 339
  Traditional Solution 339
  Comments 340

The Gyroverse
The Hidden Structure of the Universe

    Gyroverse Explanation 341
CHAPTER 12 – INSTANT MESSAGING 342
    Introduction 342
    Description of Experiment 342
    Conclusion 347
BIBLIOGRAPHY 349
INDEX 381
ABOUT THE AUTHOR 399

Donald Wortzman

Part I - The Gyroverse Model

CHAPTER 1 - INTRODUCTION

Too many confounding behaviors of the Universe are not supported by our three-dimensional view. For a plausible explanation, a new physical construct of the universe is needed. Take, for example, the behavior of "quantum entanglement," the phenomena where particles that once interact, retain an uncanny awareness of each other, even after being separated by many miles. This aberration, called by Albert Einstein, "spooky action at a distance," can be explained if these particles are, in reality, very close to each other. Another confounding mystery, which Isaac Newton was first to identify, is the equivalence of inertial, gravitational, and energy mass. These three seemingly unrelated phenomena should not have the same mass value. In spite of this, they do, because they all have a common origin that is not apparent from the three-dimensional view of the universe. Many other unexplained natural occurrences will be described in this treatise, each of which further helps to reveal the true nature of the universe. Intuition, common sense, and logic can no longer be relied upon because the physical structure of the universe is very different from its appearance.

A new gyroverse model of the universe explains most mysterious anomalies. In this model, the universe is composed of twelve dimensions, which are related in shape to a hyper-cylinder with proportions the size of an atom. All matter is wrapped around the twelve-dimensional hyper-cylinder moving on this helix at the speed of light. The immense inertia caused by this motion constrains all movements to this three-dimensional subset. Light and all other reference measurements follow the same rotational path. All bending is confined to dimensions other than the three-dimensional Euclidean subset, so that the rolled-up nature of the universe is hidden, giving the universe its conspicuous three-dimensional flat appearance. Distances, as far as light-years away, in three dimensions are reduced to atomic distances in the full twelve-dimensional space. Though it seems implausible for the entire universe to fit into such a seemingly small manifold, it can do so with room to spare. This is because in high dimensional spaces, the distances can be kept small, still allowing the volume inside to be immense. While the Euclidean subspace that we reside in appears as three-dimensional, it will be shown to be a four-dimensional hyperspace embedded into a twelve-dimensional space. Nonetheless, the twelve-dimensional construct is not merely a convenience for containing the four-dimensional subset, but is essential in its own right. Certain characteristics only make sense from the twelve-dimensional vantage point. Others are grasped more easily using the four-dimensional or even the three-dimensional perspective.

While introducing more space dimensions may seem to complicate the issue, several theories, including string theory and Kaluza-Klein theory, are famous for having introduced the notion of a universe with more than three space dimensions. However, both theories retain the current three dimensions and only use these small extra dimensions to slightly augment the original three. Thus, these are merely mathematical devices that the proponents hope will clarify the mathematics describing the physical universe, but it will still leave most questions unresolved. This physical construct, the gyroverse model, takes the next logical step.

The intention here is to validate all currently known physics in terms of this model, with emphasis on those areas where this model most affects present accepted thinking. This theory should be reviewed, not from the perspective of whether each detail is perfect, but whether this model is on the right track and can ultimately better explain the workings of the universe. It has been said that an advantage of language is that ideas can be exchanged without requiring exactness. If normal language were too precise, like mathematics, all statements would be black or white with no shades of gray. Some vagueness enables new ideas to be refined, a bit at a time. While complete rigor is ultimately needed, it often gets in the way if insisted upon too early.

This book is separated into three parts. Part I contains the theory describing the gyroverse model and comprises seven chapters. Part II is an overview of the prevailing theory, consisting of three chapters. Part III has two chapters, with thought-experiments of special importance. Because of the many new ideas, a summary of each chapter is included at the end of this chapter to introduce the material gradually.

The book is not too mathematical, but contains some only when it is believed to be of value. The presentation can be understood, even if the math is skipped. Those wanting to enhance their familiarity with these subjects before tackling the theory should read the background chapters in Part II, an overview of current theory. To make different sections of the book understandable in isolation, some explanations are repeated to reduce the need for skipping around the book. Besides, sometimes repeating information, but saying it in a different way enhances understanding.

This model is described much as an automotive engineer might explain how a car engine works, emphasizing the basic principles involved, and not the precise equations. The real universe is an extremely sophisticated mechanism, very different from its appearance, and cannot be understood with mathematics alone. Bertrand Russell underscored this point when he observed "Physics is mathematical not because we know so much about the physical world, but because we know so little; it is only its mathematical properties that we can discover." However, most equations are approximations of reality, and are accurate only within certain bounds. For example, Newtonian laws are incorrect for high speeds, relativity is invalid for atomic structures, and quantum mechanics is wrong for large masses. The point is that none of these theories matches the physics exactly, and the equations cover a larger domain than the physics would justify. The same expressions are often used in completely different disciplines, and do not know for what they are being used. Applying them at their fringes and concluding that time can go backwards, or kinetic energy can be negative is extremely problematic. It must be remembered that equations do not think; we must do all the thinking.

This theory aims to correct the in-vogue tendency of straying from the physical with mathematics. From a practical point of view, a few changes to current physics equations are required, but very significant changes to the understanding and explanation of all physics. Inflation theory is gone and the universe expansion mechanism is completely revamped. The reason for the lightspeed limitation is explained. Another significant change is in the elimination of "non-simultaneity," a strange byproduct of special relativity. In quantum physics, the causes of quantum entanglement, duality, and particle spin are given. The theory explains the seven forces of nature, including the four familiar ones, all derived from the same underlying mechanism.

Finally, because the idea of more physical space dimensions is a very significant departure from our present-day idea of the universe, it is explained gradually building on familiar ideas. They are developed in three-dimensional spaces, where visualization is manageable and extended to higher dimensional space where the notions are needed but visualization is difficult.

CHAPTER-BY-CHAPTER OVERVIEW

The book includes a chapter-by-chapter overview. Including this in a book is unorthodox. However, this material is so unique, influencing all current physics thinking that it better prepares one for the more detailed full presentation. Do not be concerned with not following the entire overview, it only meant to acquaint the reader with the basic ideas. An old saw for making something understandable says: "First, tell them what you will tell them. Then tell them. Finally, tell them what you told them." A worthwhile idea is to re-read this chapter after Part I is finished, since it will be much more readable, and will provide an excellent summary of the book.

RE: CHAPTER 1 - INTRODUCTION

The first chapter sets the stage for taking a fresh look at the current accepted embodiment of the universe, and its effect on the two major physics theories, quantum mechanics and relativity. Two, of many, unusual behaviors of nature are discussed here. A brief description of the structure of the gyroverse model is presented.

This book is separated into three parts. Part I contains the gyroverse theory, Part II, an overview of present physics theories, and the last part discusses two thought-experiments of special importance. A chapter-by-chapter overview of the book is presented.

RE: CHAPTER 2 – THE PROBLEMS

More than a dozen anomalies are discussed here with compelling evidence that the Euclidean appearance of the universe is not its real physical structure. These include quantum entanglement, the phenomenon in which particles that interact once, remain coordinated, even after being separated by miles. Another is the equivalence of inertial, gravitational and energy mass. These three seemingly unrelated phenomena should not have the same mass value, but do.

Several of these anomalies offer ample evidence that something is amiss, but taken together they give a strong indication of the actual structure of the universe. In the gyroverse model, the universe consists of twelve dimensions that enable all parts of the universe to be only atomic distances from all other parts of the universe. It is this surprising construction of the universe that is the underlying reason so much of the physics appears illogical.

RE: CHAPTER 3 - THE MODEL BASICS

The major tenet of this model is that all real distances within the universe are of atomic proportions. Distances as far as light-years away in three dimensions are reduced to atomic distances in the full twelve-dimensional space. This is accomplished by rolling up Euclidean space into a twelve-dimensional manifold, each axis being rolled up in a tightly wound helix. In this chapter, a systematic progression of arguments to justify this claim is presented. Several examples are given to convince the reader that the whole universe can fit into a tiny many-dimensional space, and still look as it does.

Perhaps the hardest thing to grasp is where these extra dimensions might be. To appreciate the difficulty, an analogy to Edwin Abbott's classic <u>Flatland</u>, about a two-dimensional society is presented. In this book, the narrator A. Square tries to reveal his experience with a three-dimensional sphere that he met, to other Flatlanders. He can only point in the four surface directions, and not up or down, making it very difficult to describe the third dimension that no one has seen.

Introducing additional space dimensions might appear to complicate the issue rather than explain it. Nonetheless, string theory, which is now widely accepted in the physics community, has made the idea of a universe with more than three dimensions palatable. String theory retains the current three infinite dimensions and only uses these extra dimensions to supplement the original three. The gyroverse model takes a different tack by showing how the three infinite dimensions can fit into a small many-dimensional manifold.

RE: CHAPTER 4 - THE GYROVERSE MODEL

An important objective of the modeling method is to make this twelve-dimensional manifold comprehensible. Recognizing the symmetry of Euclidean space, a three-dimensional representation of one axis is developed and duplicated for the other two axes. A line wrapped on a hyper-cylinder of atomic proportions, pitched forward forming a tightly wound helix, represents each direction in Euclidean space. Added to this is a hidden fourth direction, represented by a helix wound identically to the other three that in special relativity is taken to be the time dimension. All matter is circulating an advancing helix for the fourth direction at the speed of light. Thus, the entire twelve-dimensional space model consists of four identical copies of a three-dimensional cylindrical helix that represents each direction in Euclidean space.

Each axis, x, y, and z, is wound within its own separate three-dimensional space. From the x, y, and z perspective, the absence of bending between these separate three-dimensional space boundaries, makes the space appear Euclidean. All bending is done into space dimensions that are not perceptible to us. Because matter has zero thickness in these dimensions, the rolled-up objects are not stressed and can be bent this way, not being noticed. Though each axis is wound tightly in its own subset, the universe still appears Euclidean.

Circulating matter in the fourth direction creates a powerful multidimensional gyroscopic-like mechanism, called the gyrohelix, which keeps the universe stabilized, and provides all of the universes energy. The gyrohelix is held together with small particles called graveons, which permeate space. Matter has little, if any, intrinsic mass-like quality; the mass is generated by the gyrohelix mechanism. This gyroscopic action creates and amplifies inertia by 30 orders of magnitude, starting from almost nothing.

Gravity is not pulling on matter, but graveons are pushing matter together causing gravitational attraction. This pushing action impedes matter from accelerating but allows it to move freely at a constant velocity, giving it its inertial quality. The rest energy of matter is attributed to the kinetic energy of matter traveling at lightspeed in the fourth direction, and not due to matter being intrinsically equivalent to energy.

Clock time changes because of the inertial mass change, but real time does not. Special relativity, by means of the Lorentz transformation gives the formulae, but not the mechanism, for distance's apparent contraction, time clock slowdown, and mass value increase when traveling at high speed. The Lorentz transformation's gamma factor, the crux of the transformation, will be derived from gyroverse basics. However, its interpretation has crucial differences from special relativity.

RE: CHAPTER 5 – RELATIVITY IMPLICATIONS

This chapter is concerned with those aspects of the gyroverse that deal with the physics addressed by special and general relativity. It begins with a discussion of non-simultaneity. Non-simultaneity is one of the strangest aspects of special relativity finessed by the gyroverse model. Non-simultaneity occurs when one observer recognizes two events in another inertial frame as happening at once, while another observer, distant from the first, will record supposedly completely different times for the event pair, possibly years apart. In this theory, all observers will agree on the timing of all events, after correcting for their different clock rates. The reason lightspeed is the same, and limited to c in different inertial frames is explained. Three important gyroverse terms, yaw, pitch, and roll are discussed. Yaw is the direction the gyrohelix faces; pitch is the advance for each revolution of the gyrohelix; and roll is the angular velocity of the gyrohelix. These are the parameters that were set once the universe accelerated expansion subsided.

Several equation differences the gyrohelix and special relativity are discussed that significantly influence the physics. One concerning the Lorentz transformation is that for this theory only one principal frame exist; any other frame has an elevated gamma, causing apparent distance shortening, mass increasing, and time clock slowing as viewed from the principal frame. This gamma change cannot be easily detected, because the standards that they are compared against experience the same changes. However, distances don't actually shorten, only appear so. In fact, each views a shortening of lengths in the other's inertial frame in the motion direction, but it occurs for a different related reason. In addition, the velocity transformation (velocity addition) formula is significantly different from the special relativity counterpart. While the special relativity formula was presumably verified by the Fizeau experiment, that result is disputed.

RE: CHAPTER 6 - QUANTUM IMPLICATIONS

As in the previous relativity chapter, this chapter addresses those aspects of the theory that influence quantum physics. The differences with present-day theory are not numerical or mathematical. However, aspects of quantum theory that were previously unexplained can be now understood. Gravity is a very weak, long-range force in Euclidean space, but a very powerful short-range force in twelve-dimensional space. The force is caused by graveons uniformly traversing twelve-dimensional space. This modality is also responsible for the strong force that holds the atomic nucleus together, and the weak force that governs the particle decay, for example a neutron into a proton and electron (plus an antineutrino). Another facet of this force is holding entangled particles together. Entangled particles can be far apart in Euclidean space, but remain exceedingly close in twelve-dimensional space. They maintain an awareness of each other even after being separated by miles. The mechanism for providing these seemingly different forces with one basic force is explained in detail.

The electromagnetic force, which repels like-charged particles and attracts unlike-charged particles, is also explained. This dual path mechanism defies explanation in a three-dimensional Euclidean space. The vehicle for this force is the photon that indirectly gets its energy from graveons impinging on it. This dual path mechanism, when applied to a particle is the basis for its antiparticle. When both meet, even with little apparent relative motion, they crash at twice the speed of light causing the annihilation of each, and an immense energy release.

Duality, in which particles have both wavelike and particle-like properties, is described as well. Small particles can take shortcuts in twelve-dimensional space, remaining small, and compact, but dispersed over Euclidean space. It will be explained, for example, how a photon or electron can actually pass through both slits simultaneously in the classic two-slit experiment.

Fermi particles such as electrons, protons, and neutrons have a propensity for keeping their distance from each other. Bose particles, such as photons and graveons, which transmit the forces of nature, have an affinity for bundling. The mechanism for each is explained. Atomic structure, including the root cause of atomic particle spin, is discussed within this framework.

RE: CHAPTER 7 - COSMOLOGY IMPLICATIONS

The big bang theory is the predominant explanation for the origin of the universe. This theory was proposed when it was discovered that the universe was expanding. The reasoning was that if it were always expanding it must have been tiny some time in the past. At first, the two aspects of general relativity were sufficient to explain the universe's evolution. These are the traditional inertial motion of close celestial bodies, and the general expansion caused by the stretching of empty space that does not involve inertia. However, the flatness and horizon anomalies made this explanation problematic. Subsequently, an inflation theory was proposed to resolve these observed problems. Inflation, as defined, only existed for a fraction of a second during the very beginning of the universe creation and then dissipated. More recently, it was discovered that not only is the universe expanding, but it appears that the expansion is accelerating. This is another big surprise because this was inconsistent with expectations. It was assumed that gravity would work to slow the expansion, even if it were not great enough to stop it. Thus, a gravitational repulsion mechanism had to be added to the theory. In effect, four cosmology theories were stitched together that barely coexist. However, this treatise will show that the expansion is not accelerating, but just that light from distant galaxies take a curved path in reaching us, making the distant traveled longer than assumed.

The facts surrounding the big bang theory are reinterpreted to be consistent with the gyroverse model, while combining all the current loosely coupled physics theories. Neither Doppler shift, nor the stretching of space, causes the red shift of light from distant bodies. Celestial motion involves real movement of matter in the fourth direction on an expanding 3-sphere, giving the appearance of expanding space. Unique to this motion-direction expansion is that it is inertial, but required 28 orders of magnitude less energy to achieve it. This force causes acceleration without a corresponding change in yaw, increasing the inertial mass with very little expenditure of energy. Familiar energy and mass are not intrinsic properties of matter, but are generated by the gyroverse mechanism. The universe is much less substantial from the outside than it is from the inside.

Also presented are illustrations showing how the universe looks today and how it looked at its inception. While the entire universe was originally subatomic in the four-dimensional subset, it was not compressed, since the entire universe could fit comfortably into that space. Graveons that emanated from all matter expanded the universe as a whole, while it still caused nearby matter to congeal under the force of gravity. The force of expansion subsided as the universe enlarged, but did not completely vanish. Thus, the big bang is better characterized as a big push that still barely persists. Compactness of the universe initially allowed the subatomic particle to entangle and remain connected during expansion, keeping the universe's energy distribution uniform, before neutral atoms formed.

The gyroverse model predicts a slightly older universe, because it initially expanded slower, giving large celestial bodies enough time to form. In addition, the expansion of the universe is not in the same dimension as the gravitational force that formed celestial bodies. The universe is much smaller than present orthodox theory assumes. At a distance of 13.5 light-years, the present maximum look-back, just about all of the matter in the universe is observable.

From the angular velocity of revolving galaxies and the amount of visible matter, the centrifugal force is too large for gravity to hold the galaxies together. To reconcile the dilemma, it was proposed that the universe contains much more unseen dark matter than what is visible to supply the extra needed gravitational force to hold the galaxies together. An alternate, much more plausible theory, with acronym MOND, posits that inertial mass is much smaller at ultra low accelerations, the condition at the outer margins of the revolving galaxies. While this would resolve the missing matter dilemma, without the need for dark matter, it contradicts general relativity; however, it reinforces the gyroverse theory, and is the most probable solution to the missing mass problem.

Another problem considered is stellar aberration, which causes a star's apparent position to change if the observer, but not the star is moving. This is a blatant violation of special relativity, which posits that velocity is not absolute; a star's position in the sky should only depend on the relative velocity between both. In contrast, it is especially consistent with the gyroverse theory.

Also discussed is Wheeler's delayed choice experiment, where light can take two paths to a detector. Incredulously, by modifying the detection configuration just prior to the photons reaching the detector, the delayed choice hypothesis infers that photons can go backward in time, and take the alternate path. While this experiment has been done in the laboratory, the original proposal had light emanating from quasars hidden behind galaxies, taking two paths around it. Then, the back-up time would be millions of years. The gyroverse theory leads to a much simpler and more plausible explanation.

In addition, several irregularities that have stymied current theory are explained. One of them is the Pioneers 10 and 11 probes mystery, where both deep space shots are moving away from the sun slower than expected. Another is the Eclipse dilemma, where the pull on earth dips slightly as the sun and moon enter into an eclipse.

RE: CHAPTER 8 - RELATIVITY OVERVIEW

This is the first of three chapters that are overviews of existing accepted physics. It has been placed after the theory so that readers not recently acquainted with the traditional presentations of quantum mechanics, relativity, and cosmology can have this information available before they read the gyroverse model. Besides presenting the highlights of these theories, a historical perspective of each is also presented. Readers familiar with the subject may also find them helpful, because the material covered is more directed to the gyroverse theory.

The chapter starts with Sir Isaac Newton who gave physics its mathematical foundation and initiated it as a separate discipline in science. Discussed are his three laws of motion and his discovery of the laws governing gravitational attraction that he published in his famous book <u>Principia</u>. The Galilean transformation is a procedure of changing the reference between two inertial platforms, moving relative to each other. Newton's laws of motion are covariant with respect to this transformation, working the same on any inertial platform.

His theory prevailed for more than 200 years before a serious crack in the theory developed from experiments by Michelson-Morley and Fizeau, in the late 1800's. This showed that the round trip speed of light was the same in all directions on Earth. Attempts were made to reconcile these experimental results by postulating the presence of an undetectable aether gas, permeating all space. Aether gas was given all the necessary characteristics to explain the Michelson-Morley and Fizeau results within Galilean transformation, but it failed. Maxwell's electromagnetic equations are not covariant to the Galilean transformation. The Lorenz transformation, originally postulated to allow these Maxwell's equations to be covariant to moving inertial frames, later became the heart of Einstein's special theory of relativity. The major characteristics of special relativity are that objects in motion experience mass increasing, distance contraction, and time dilation. General relativity extends this understanding for objects during acceleration or under the influence of gravity.

The tipping point for the embracing of General relativity was when the bending of light skimming the sun, during an eclipse, was measured to be precisely what relativity predicted. Newtonian physics only accounted for half that amount. The icing on the cake was the correct accounting for the precession of the perihelion of Mercury. However, now both of these phenomena can be explained without invoking relativity theory. The calculation for the bending of light is included earlier in this treatise.

RE: CHAPTER 9 - QUANTUM OVERVIEW

The Greek philosopher, Aristotle, believed light originated from the eye directed outward toward the object, much as a blind man uses a cane. It took more than a thousand years before an Egyptian philosopher showed that light was reflected from objects to the eye. Newton believed that light was composed of tiny particles, or corpuscles, emitted by luminous bodies. About the same time, Christiaan Huygens developed a theory that explained light to be a wave phenomenon. While the evidence for the corpuscular theory was not as strong, the stature of Newton kept that theory alive. This prevailed until Thomas Young and Augustin Fresnel did diffraction and interference experiments, which could only be explained from a wave theory of light. Wave theory was further cemented with Maxwell's wave equations that merged light with electromagnetic theory. In the early nineteen hundreds, Einstein presented a paper on the photoelectric effect, postulating light's particle-like qualities, subsequently called the photon. This was the origin of the duality principle that light was both a wave and a particle.

Niels Bohr, a protégé of Rutherford, used his atomic model, the quantum ideas of Einstein and Planck, and proposed that the electrons revolve around the nucleus in a very limited set of circular orbits with discrete energy levels. The model initially explained all observations, but later conflicted with experimentation that were more precise.

Louis de Broglie developed the notion that matter also had wavelike characteristics. Erwin Schrödinger analytically extended the idea, formulating the quantum theory of wave mechanics. This describes the behavior of atomic particles as waves, extending the particle-wave duality of light to matter.

One of the most instructive thought-experiments devised is known as Schrödinger's cat. What Schrödinger did was to combine subatomic phenomena in which we do not have any personal experiences and can accept indeterminacy, with a cat, where we have everyday experiences to rely on, to show an absurd nature of quantum theory.

Experiments, leading up to, and verifying quantum entanglement and the testing of Bell's inequality is discussed in detail. This began as an ongoing debate between Niels Bohr, the defender of quantum theory, and Einstein who thought that there were, as he described, "hidden variables" that would clear up the strangeness of quantum theory. He devised a thought-experiment called the EPR paradox to prove his case. History has so far judged Bohr arguments to prevail.

The standard model with its various atomic particles and particle families is presented. Discussed are the differences between bosons and fermions. Quarks, which in various combinations make up protons and neutrons, are defined. Structure of the atom is also discussed.

The four forces in nature are defined. Gravity and the electromagnet force are forces that are commonly experienced. The strong force holds the atom nucleus together, and the weak force regulates a variety of nuclear decay processes, such as a neutron into a proton, electron and antineutrino. A theory that combines the electromagnetic and weak forces into a single electro-weak force has been accepted by the physics mainstream.

Quantum electrodynamics, or QED for short, is the interaction between photons and electrons. Most familiar everyday phenomena involve this type of interaction, for example, chemical reactions, biological processes, and the characteristics of matter. Quantum chromodynamics (QCD) describes the interaction between quarks and gluons in the atom's nucleus.

The goal of modern theoretical physics has been to find a unified description of the universe, a single theory to explain all the atomic particles and the four forces of nature. Kaluza-Klein theory was an early attempt to unify physics theories, by adding a fifth hidden dimension to the universe. String theory is the modern-day attempt to create this theory of everything (TOE) as it is sometimes called. It extends the Kaluza-Klein idea, defining a universe that has many hidden dimensions.

RE: CHAPTER 10 - COSMOLOGY OVERVIEW

The Greek astronomers were the source for the cosmological concepts and theories until Copernicus. The central theme of ancient Greek cosmology was the geocentric theory that claimed that Earth is at rest and the center for all other motion in the universe.

Nicolas Copernicus, a Polish astronomer who lived between 1473-1543, degraded the importance of Earth when he introduced the heliocentric model that had the sun at the center with Earth revolving around it. He claimed that all the planets, not only Earth, moved in orbits around the sun. This explanation was extended and enhanced by Galileo who was first to use telescopes and Kepler who added mathematical accuracy.

In his book <u>Principia</u>, completed in 1687, Newton proposed his three laws of motion that still stands today for most practical problems. In addition, he discovered the universal law of gravitation, a formula for the gravitational attraction between two massive bodies. It extended Kepler's idea by describing the motion of the solar system in analytical terms. He was a co-inventor of calculus, most important for accurately calculating the motion of celestial bodies and many other scientific investigations.

Discussed here are the three most important techniques for estimating the distance to celestial bodies. These are parallax shift, a form of triangulation; standard candles, stars whose luminosity can be estimated; and red shift of light from celestial bodies.

Hubble, using very powerful telescopes, was first to discern that the universe is expanding uniformly. This caused the big bang theory to replace the steady state model, as the explanation of the universe's evolution. The inflation mechanism was incorporated into the theory to explain the flatness and horizon dilemmas. More recently, there has been measurements to suggest that the universe's expansion is accelerating, but an alternate explanation is presented.

RE: CHAPTER 11 - TRIPLET PARADOX

The triplet paradox is an extension of the famous twin paradox thought-experiment introduced by Einstein. It deals with three identical triplets, Tinker, Evers, and Chance. Tinker and Evers live on Earth, and Chance is selling real estate on the planet, Soil, 4 light-years away. Tinker plans to visit his brother. This thought scenario highlights the strangeness of non-simultaneity introduced with special relativity, but this paradox is better explained within the gyroverse theory.

Re: Chapter 12 – Instant Messaging

This thought-experiment uses ideas akin to the delayed choice phenomenon and quantum entanglement, to effect communication faster than the lightspeed. Einstein used the speed of light limitation to argue wrongly that entangled particles are not really influencing each other, but instead hidden variables create the impression that they do. While it can be argued that quantum entanglement is already an example of communication faster than the speed of light, as Einstein's argument would suggest, the apparatus disclosed in this treatise removes any doubt.

CHAPTER 2 - THE PROBLEM

INTRODUCTION

Science has uncovered the complex secrets of the universe, from the smallest of particles to the outer reaches of space, billions of light-years away. The two most profound theories developed in the twentieth century to explain these behaviors are quantum mechanics and relativity. Quantum mechanics describes the behavior of atomic particles, while relativity does the same for larger matter. For some situations, their application does overlap.

Both theories, in their realm of applicability, give extremely accurate results, but do not join seamlessly. In fact, they are incompatible, and are rarely applied simultaneously to the same problem. Relativity is deterministic, in that given the present state of all matter, its position and velocity will determine, in principle, the future state of all matter. For example, astronomers can predict moon and sun eclipses well into the future by knowing the current state of all nearby celestial bodies. The situation with quantum theory, which applies best to atomic size particles, is more limited. Only the probability of a particle's state, its position and velocity, can be known at any given time. By accurately measuring one, the other becomes vague. Thus, even if one knew the existing state of quantum particles, their future state could not be accurately predicted, even in principle. Although, for most practical situations, this limit is not a problem, it is still disconcerting that nature has placed boundaries on the extent of our knowledge. Werner Heisenberg formulated this limitation to knowledge, the

Uncertainty Principle, while working at the Institute for Theoretical Physics in Copenhagen.

Left in the wake of this theory is a host of basic operations of nature that are not understood, one dating back to Isaac Newton. Rather than being dissatisfied with this lack of knowledge, physicists have elevated these anomalies to "axioms of nature," not requiring explanation. In effect, nature works mysteriously, and all one can know is the mathematics for getting correct answers. Physics has become metaphysical; reality does not exist until it is observed. This hopelessness of ever understanding the workings of nature permeates the scientific community. The great physicist Paul Dirac made this point when he said, "The only object of theoretical physics is to calculate results that can be compared with experiment. It is quite unnecessary that any satisfying description of the whole course of the phenomenon be given." This is the exact antithesis of this theory, which is not about calculation, but about understanding how nature works.

It seems that scientific investigations have bypassed the physics' mysteries that past masters tried and could not answer, and instead are investigating ever smaller and less accessible atomic particles on the one hand, and delving deeper into the furthest reaches of the universe on the other hand. Mistakenly, they believe that the answers to classical mysteries will be a natural byproduct of this strategy. Richard Feynman, perhaps the most renowned physicists in modern times, further summed it up in his book QED. After explaining some unusual aspect of quantum physics' famous double-slit experiment, he states the following: "I have pointed out these things because the more you see how strangely nature behaves, the harder it is to make a model that explains how even the simplest phenomena actually work. So theoretical physics has given up on that." In that same spirit, he is also credited with saying, "I think I can safely say that no one understands quantum mechanics."

He is such an important voice in modern physics that it is negligent not to digress and include a bit about him, before going on. Richard Feynman, a Renaissance man, and a modern icon of physics, was revered by students and his colleagues alike. He was born in New York in 1918, and died in 1988. He did his undergraduate work at MIT, received his Ph.D. from Princeton, and worked on the Manhattan Project (development of the Atomic Bomb) during the Second World War. Feynman was a professor of Theoretical Physics, first at Cornell University, but for most of his career at Caltech. His knack for simplifying complex physics concepts was only possible by having the deepest understanding of the subject. Feynman achieved national prominence by his participation in the investigation of the space shuttle Challenger disaster. He did a simple test that displayed the brittleness of the O-ring seals after removing them from a glass of ice-cold water. Much of his scientific wisdom is chronicled in a three-volume set, Lectures on Physics, derived from a series of classes he gave at Caltech. In 1965, he received the Nobel Prize in physics (shared with Julian Schwinger, and Sin Itiro Tomonago) for his work on quantum electrodynamics (QED), a simplified diagrammatic approach to solving otherwise complex problems.

Nevertheless, in spite of Feynman's outlook, this treatise presents a model of the universe that explains how nature works. Unlike pure mathematics, physics has two sides, the mathematical, and the physical. For the past 100 years, the physical had been neglected, since logical explanation in terms of the Euclidean model of the universe is often not possible. Therefore, the reason that intuition, common sense, and logic do not apply is that the real universe is very different from its Euclidean appearance.

In every field, the establishment, with tacit complacency of the populace, has complete control over the acceptance of new ideas. On the one hand, this establishment is needed to oversee the discipline, but on the other, they prevent new thought from getting an impartial review. A personal experience really drove this point home.

One day, in New York City, I decided to visit the Guggenheim Museum. For those not familiar with this museum designed by Frank Lloyd Wright, it is known as much for its architecture as for the modern art exhibits within. The building, on fashionable Fifth Avenue resembles an inverted truncated cone. Within the building, a spiral ramp hugs the outside wall, where the artwork is placed. This arrangement is especially conducive for exhibiting artwork best viewed in some special order, for example, chronologically. However, its small ratio of wall space to interior room is at odds with the high rents this location demands. After paying an entrance fee, my wife and I started up the ramp. All that could be seen at the first location was a blank white wall. In fact, all along the ramp there were no paintings, but just blank walls. Thinking, they must have made changes to the museum since last there, we went back down the ramp, and stopped at the front desk to ask where all the paintings were. The attendant said that they removed the art to paint the building's interior, and thought this was a good opportunity to feature the building architecture, which can better be appreciated without any distraction from the artwork. You mean, "We paid just to see blank walls." He did not seem to appreciate our annoyance. If one bought this book filled with blank pages accompanied by an interesting cover jacket, I probably would be accused of cheating, no matter how innovated the explanation. However, in the art world, the insiders can decide that blank walls or even a painting of a Campbell soup can is art. The idea that the in-group can act so outrageously and not face any serious criticism was mind-blowing.

New ideas are embraced gradually. Max Planck observed that science never accepts a new theory until enough of the old-guard opposition die off. The forums where this work might be presented are by invitation only and limited to the current paradigms. The acceptance committees come from the same pool of experts. Therefore, new ideas cannot easily penetrate this wall until an overwhelming consensus is built within this community. The Internet, potentially, can break this stranglehold. This work was displayed on the Internet at its own Web site. It was registered with the major search engines and referenced by several Web sites that specialize in new physics theories. Steady streams of people viewed it--something not conceivable even just a few years ago.

Physicists are fond of saying that a new theory is not justified, unless it predicts things that cannot be predicted from existing theory, while not contradicting contemporary proven physics. While that normally makes good sense, the list of anomalies that are not explained by present-day theories is so large that clearly they are not complete, even though they are numerically accurate where they apply. Unfortunately, a simple test to prove the gyroverse theory cannot be provided, but it makes many "postdictions," explaining most physics enigmas. In addition, it presents many links between the fundamental aspects of physics, greatly reducing the number of independent variables.

Clearly, the gyroverse model must satisfy several criteria. First, it must be consistent with existing experimentally proven physics. Any variances must be minor enough not to be measurable. Second, it must explain how it can be so very different, and still enable the universe to maintain its present appearance. Finally, it must resolve the enigmas described. The more anomalies explained by the model, the more confidence is justified for the value of the model. The gyroverse meets this test, because the list of anomalies it explains has grown large.

One disadvantage of describing many enigmas is that in some minds it lessens the importance of each; "the more is less syndrome." However, several of these alone provide ample evidence that something is amiss with our assumption that the physical structure of the universe is as it appears. The importance of considering all of them is that together they suggest the nature of the solution. As the model is developed here, it will become more apparent how the various enigmas were suggestive of particular aspect of the model.

One overriding difficulty for understanding something new is unintentionally filtering it through the established "belief system." That process can be detrimental for understanding a new idea. Each time it takes conscious effort to listen--understand before judging. Many problems in physics that this theory addresses have been around for well over 50 years, one for several hundred years. If understanding them only required a slight alteration of conventional wisdom, they would have been understood many years ago. Often someone outside the mainstream is needed to scrutinize physics, going back to basics.

An important principle that underlies all scientific inquiry is to select the simplest alternative among competing theories. The 14th century English philosopher and theologian, William of Occam, popularized this notion known as Occam's razor. In practical terms, one should always choose the simplest explanation among competing scientific explanations or theories, with the fewest "leaps-of-faith." Trim away all the constructs that are not needed. Clearly, this process is subjective and not always easy to apply, but having this ideal is useful. In this treatise, you are being asked to accept a new model of the universe. Unfortunately, just as the current theories cannot be proven, a new theory also cannot be proven. You must decide whether the proposed theory is better than the existing theory with subjective criteria, such as Occam's razor test.

It is interesting how much intelligence can be gleaned from paltry information if it is examined from a different perspective. Suppose someone was presented with three coins of unknown origin, but wanted to learn as much about them as possible. Assume that the first coin is the most valuable, it takes 2.00 of the second coin to exchange for the first coin, but it takes 2.37 of the third coin to exchange for the first coin. What could be inferred about the three coins, besides the obvious, that the first coin is the most valuable and the third coin is the least valuable? If one thinks about it a while, he will also realize that the first two coins come from the same country (or a country that pegs their currency to the other), and the third coin comes from a different country. What country would peg two of their coins at a 2.37 for 1 exchange ratio? Imagine the United States issuing a new larger denomination coin, where it takes 2.37 half dollars to exchange for one of these new coins. A math major would be required just to make correct change. On the other hand, what is the probability of two different countries' currency being in an exact 2 to 1 exchange ratio? It is possible, but highly unlikely. The point of this exercise is that by Occam's razor, without knowing for sure the countries of origin for the three coins, logic alone could glean enough information to make informed predictions. Both assumptions; that a single country would have currency not in simple ratios or two different country's currencies would be in simple ratios, needs to be eliminated, because they are both improbable--do not believe in coincidences. In the set of anomalies that will follow, this type of reasoning is applied often.

RELATIVITY ENIGMAS

ENERGY-MATTER EQUIVALENCE ($E = mc^2$)

If one thing every layperson knows about special relativity, it is $E=mc^2$, where E is energy, m is mass, and c is the speed of light. It is interesting to consider the significance of the different parameters in this equation. It is clear why a measure of mass enters linearly into this relationship. For example, if one has twice as much matter, expecting that it contains twice as much energy is reasonable, just as twice as much gasoline will take your car twice as far. Nevertheless, why would the m in mc^2 be identical to the m of inertia? One would expect it to be approximately the same, but not identical to another unrelated measure of the quantity of matter. The fact that they are identical implies a physical relationship between the two. What is even less obvious is what the speed of light has to do with how much energy into which a chunk of matter converts. Einstein proved that the rest energy of matter is mc^2, so the issue is not whether the equation is correct--it is. In fact, he proved it two different ways, one using the result of special relativity and the other proof was independent of relativity. The issue here is why nature behaves this way. One would have expected that to convert matter to energy, the constant of proportionality would be a new parameter of nature that Einstein discovered. If a certain parameter of nature appears in two seemingly unrelated phenomena, then a hidden physical relationship between the two phenomena, or an overwhelmingly improbable coincidence, must exist. Thus, the fact that matter, supposedly a condensed form of energy, has the speed of light as a factor for converting it to energy is very significant. What is even more startling is that the constant is not just c, but c^2, since mc^2 is in the exact form of the kinetic energy equation, the energy of motion. In fact, if all matter were really moving at the speed of light, the energy of motion would be exactly mc^2, where m is the inertial mass and c is the speed of light. This suggests that the matter might be getting its energy from motion and not from the matter itself changing into energy, which would also explain why energy mass and inertial mass were identical. After all, if all matter moved together at the speed of light, only their relative motion would be observed. When looking out a train window at another train on a neighboring track moving in the same direction, one gets the impression that the train is standing still. This is a common occurrence in other similar situations where stationery background objects are not visible. Surprisingly, physics books do not seriously question the nature of the formula. One speculated that all the internal motions in matter just happen to add up to this amount of energy--unlikely.

However, while the casual observer would not be aware of the component of motion common to all matter, it is likely that a perceivable difference in the laws of motion when moving in the common direction as opposed to other directions would be observed. Since no difference in the laws of motion in any direction is noticed, it suggests that the common motion must be in a direction other than the three conspicuous directions in space. This is precisely what this treatise predicts. The exact nature of this direction of motion, where all matter is moving at the speed of light, will be described.

INERTIA, GRAVITY, AND ENERGY EQUIVALENCE

Isaac Newton found it puzzling that the mass of inertia in the formula $F=(m_i)a$ is the same as the mass of gravity in $W=(m_g)g$. The first mass, (m_i) is determined by how much force must be applied to matter to make it accelerate or decelerate. Out in deep space, moving matter will continue its motion until an external force is applied to change its velocity. The second kind of mass (m_g) is characterized by the object's weight on Earth. This weight is caused by the mutual gravitational attraction between two chunks of matter, Earth and the object. More weight implies greater gravitational mass. Nevertheless, while inertial mass does not involve gravity and gravitational mass does not involve acceleration, both measures of mass are identical. Newton recognized this conundrum but could not explain it. He knew enough to expect that there had to be a logical connection between these two phenomena even if he could not identify it. More recently, the matter-energy equivalence was added to this quandary. Here we have a third measure of mass that involves neither acceleration nor gravity, portraying matter as a condensed form of energy. Newton never questioned this notion of mass because it came long after his time, as a byproduct of Einstein's special relativity theory.

Ernst Mach, a notable nineteenth century physicist, theorized that all matter in the universe caused inertia. Einstein, who coined the term "Mach's Principle," agreed with Mach and postulated that they were equivalent in his general relativity theory, published in 1916. However, he, just as Mach, could not provide a rational explanation for the underlying mechanism. In fact, general relativity has a solution, even with the absence of other matter in the universe, suggesting that Mach's principle is not really a precondition for it. Einstein became less enthusiastic about Mach's principle in his later years, possibly because of this fact. It will become apparent that the gyroverse theory supports Mach's Principle, and explains the mechanism involved. Our solar system could not exist without the rest of the universe.

Nevertheless, the success of the general relativity theory is just more evidence that gravitational and inertial mass are the same, although it offers little rationale. On the surface, they are completely different phenomena. It would be equivalent to the conductivity of electricity being the same as the conductivity of heat in a metal. Although the conductivities of both are similarly dependent on the free flow of electrons, their dependencies are not identical. Newton tested many materials trying to find a difference in the gravitational and inertial mass, but could find none. Since then, tests that are more accurate were conducted, but still no differences have been measured. In fact, the mass for the energy-matter equivalence, $E=(m_e)c^2$ is also measured to be the same as the other two. Many books assume that $m_i = m_g = m_e$, often without explanation or astonishment. It could be an accident of nature that made these masses the same, or more probable, a common hidden origin. The graveon, the presumed "mediator" of the gravitational force, is the common thread behind all three phenomena.

RELATIVISTIC NON-SIMULTANEITY

According to special relativity, events in one inertial frame that happen simultaneously may not appear simultaneously in another inertial frame. Elementary logic would suggest that events that are simultaneous to one should be simultaneous to all, and that the order of events should be sacrosanct for all situations. For example, if event 1 happens simultaneously with event 2, and event 2 happens simultaneously with event 3, then common logic would suggest that event 1 occurred simultaneously with event 3. This common logic principle, "Transitivity of Equality," together with "Reflexivity and Symmetry" are the most basic tenets of all mathematics. Symmetry implies that if event 1 occurs simultaneously with event 2, then event 2 is occurring simultaneously with event 1. Reflexivity requires that each event occur simultaneously with itself. It is hard to imagine anything useful where these principles are not true. However, under special relativity, the transitivity principle does not hold when events are in relative motion with each other. It is interesting that Einstein, who called quantum entanglement "spooky action at a distance" because it violated common sense, did not recognize the strangeness of non-simultaneity. Non-simultaneity is a consequence of all the laws and constants of nature being the same on all inertial frames. In the gyroverse model, this hypothesis is not assumed, creating a slight analytical variation of the Lorentz transformation, which avoids the non-simultaneity quandary.

RELATIVISTIC DISTANCE CONTRACTION

The theory of relativity states, and experiments have verified, that if two objects race toward or away from each other, their distance apart shortens merely because they are moving relative to each other. If the relative speed approaches the speed of light, the distance between these two objects shrinks toward zero. Relativity gives the equation for the distance contraction, but it does not divulge the mechanism for obtaining this contraction. The gyroverse model describes this contraction mechanism.

GRAVITATION-ACCELERATION EQUIVALENCE

A cornerstone of general relativity is that because the gravitational force and the force causing acceleration are indistinguishable, their relativistic effect would be equivalent. The argument is that acceleration and the force of gravity are two sides of the same coin. The idea behind this is that a person in a stationery elevator on Earth, or accelerating in deep space at the gravitational acceleration, g, could not differentiate between the two situations. However, a person in an accelerating elevator, out in space, will eventually increase in speed to high enough values where relativistic effects would result, distance contraction, time dilation, etc. A person in an elevator that is at rest on Earth will feel the same forces, but will not have relativistic effects no matter how long he stays. Therefore, the effect of gravity is different from the effect of acceleration. It can be argued that because it takes looking outside the elevator to notice these effects, they don't count. More reasonably, only motion itself should be overlooked, but the effects of motion are fair game. Even one difference establishes that they are not the same and in principle distinguishable. In the gyroverse theory it will be shown that gravity and inertia have the same root cause, which makes mass the same for both. Nevertheless, they are different phenomena and have some nonidentical effects. The next section describes one, somewhat less contentious than the one cited in this section.

BENDING OF LIGHT PREDICTION

One prediction made with general relativity is that light rays passing close to the sun would be bent by an angle of deviation equal to:

$$\Delta = 4GM/c^2 r$$

Where Δ is the angle of deviation, G is the gravitational constant, M is the mass of the sun, r is the radius of the sun, and c is the speed of light.

If light were considered a particle (photon) traveling at the speed of light, the bending would have been exactly half as much. Thus, Einstein attributed half the total to normal gravitation and the other half due to warping of space near the sun.

A necessary condition for the equivalence principle to hold is that inertial mass and the gravitational mass are identical. For example, in a free-falling elevator all objects inside the elevator will free float, because of this, making it indistinguishable from an elevator floating freely in space. If an object in the elevator had a greater inertial mass than gravitational mass in a free fall, it would rise to the top of the elevator. If the opposite were true, it would fall to the ground. Either of these conditions would indicate that the elevator was not in free space, but in a free fall.

The following, a variation of the preceding, is a thought-experiment that illustrates this point. Assume that inside an elevator is a photon emitter on one inside wall that emits a fine steam of photons. On the opposite wall is a screen, which displays a dot where each photon hits it. First, bring the elevator to rest far from any massive body and record the location of the photon's marks on the screen. Secondly, repeat the measurement while the elevator is resting on a massive body the size of the sun. Thirdly, repeat the measurements while the elevator is being accelerated deep in space with a force equal to the gravitational force.

The third case is strictly Newtonian, because the only thing accelerating after the photon is emitted is the elevator, with no gravitational space distortion to consider. The second case occurs on a massive body where, according to relativity, the offset should be twice as large as the Newtonian model would predict, since both the gravitational force and the space distortion is maximized here. Therefore, in this scenario, acceleration and gravity would be distinguishable, since gravity on the sun-size celestial object supposedly would bend light twice as much.

However, all is not as it seems. In subsequent chapters it will be shown that light is not bent twice as much as Newtonian physics predicts because gravity distorts space, but because gravity attracts to a point and light is also a wave. The bending due to its wave nature tangent to a massive body is small and only gets appreciable some significant distance away. Consequently, in the second and third scenario, the bending will be the same, but only half as much as general relativity would predict.

GRAVITATIONAL FORCE DUALITY

Gravity is the most common force of nature, but it only attracts. Two accepted paradigms to explain it are presented. The first, a cornerstone of general relativity is that gravity is not a force, but is caused by the warping of space-time. This explanation also was used to help explain how the speed of gravity could be limited by the speed of light and still appear to act instantaneously. The other dominant explanation, more in tune with String theorists' attempt to unify all the forces of nature, is that this force is mediated by the graveon particle. This is an attempt to characterize the gravitational force in similar terms as the other three forces in nature, which are all "mediated" by different unique particles. Clearly, both cannot be correct, or are in need of an explanation that bridges both interpretations. Often in the same book, in different chapters, both descriptions of the gravitational force are given without any mention of the apparent inconsistency between them.

It is reminiscent of the cleric that was mediating a marital dispute, first speaking privately to the husband and then privately to his wife. The cleric's wife overheard both conversations hearing the cleric tell each, repeatedly that they were right. After the couple left, the wife confronted her cleric husband, asking how they could both be right. He thought a while and the said "you know--you're right too." Similarly, gravity is both caused by the warping of space and mediated by the graveon. Physics is so accustomed to the notion of duality that no one gets upset when another dual explanation is presented.

Donald Wortzman

Space-Time Warp

    The most common paradigm to explain gravity is the space-time warping of three-dimensional space caused by the presence of matter. This explanation was introduced with Einstein's general theory of relativity completed in 1916, 11 years after he developed the special theory of relativity. A common metaphor used to explain this behavior of a warped space-time is with a two-dimensional analogy. In this model, a heavy ball is placed in the center of a taut rubber sheet. The weight of the ball flexes the two-dimensional sheet into the third space dimension. A small ball is placed at the edge of the sheet and it of course rolls toward the larger ball. Sometimes it is given a sideways push and it spirals down to the heavier ball. This metaphor is so intuitive that one immediately feels they understand the real three-dimensional case.

    A classic comedy routine by the lesser-known comedy team of Smith and Dale illustrates this point. Dale plays a Dr. Kronkite, and Smith plays the patient, Mr. Dubious, who comes to Kronkite with a medical problem. The doctor asks Dubious if he ever had this problem before. Dubious says "yes" and Dr. Kronkite says "well—you've got it again." This is a case of circular reasoning, which was funny because it was transparent and immediately recognized for what it was. Some examples of circular reasoning are subtler.

    After closer examination of the two-dimensional rubber sheet metaphor, the question arises why the small ball rolls down. Why does it not roll up or not roll at all? The answer, of course, is that Earth's gravity underneath the rubber sheet pulls both balls down. First, it flexes the sheet due to the heavy ball being pulled down the most, and secondly, it causes the light ball to roll to the lowest point. Therefore, the metaphor that explains gravity needs gravity to make it work. The rubber sheet is superfluous in this analogy. If there was no significant gravity, as in outer space far from all other celestial bodies, the heavy ball would not flex the sheet, and the light ball would not roll down toward the more massive ball. Even on an artificially flexed sheet in outer space, the small ball would not roll toward the more massive ball. If the small ball were given a sideways push, the ball would actually spiral away from the heavy ball, not toward it. This is because the outward centrifugal force caused by inertia would exceed the inward centripetal force caused by some minute gravity. This is another example of circular reasoning, but more obscure than the Smith and Dale routine. It is like trying to explain the meaning of a word using the same word. The three-dimensional case has the same flaw; it needs gravity to explain gravity, and does not help explain gravity at all. Likewise, causing anything to move in any direction is illogical for warped space. It could only explain, at most, the nonlinear path an object might take if it were already moving.

Elusive Graveon Particle

The graveon particle is believed to be the mediator of the gravitational force, supposedly jumping back and forth between objects. However, each time it leaves an object, it would recoil. Each time it strikes an object it would push forward. Therefore, a particle jumping back and forth between two objects is a recipe for repulsion, not attraction. Physicists have come up with a different paradigm to explain this attraction by assuming graveons do not cause the movement directly, but merely bring the message from one object to the other, instructing them to attract each other. This process is described by the metaphor, "mediated by exchanging graveons."

Putting aside that no physical mechanism could explain this, the particle has never been seen. Nevertheless, its existence is still predicted by different theories. The reason, it is argued that the graveon particle has eluded observation, is that gravitational force is only 10^{-38} as big as the electromagnetic force, and can easily evade detection. For example, a comb, brushed through the hair acquiring static charge could pick up a piece of paper defying the gravitational force generated by the entire Earth, which is pulling the paper down. While, at first, this reason appears logical, it is really a specious argument. Although it takes the entire Earth to generate the force (weight) that holds a piece of paper down and a small comb to pick the paper up, the two opposing forces on the piece of paper itself are comparable with each other. It does not make a difference how divergent the origins of the forces are, since they all converge on the piece of paper. If the measurement is made near the paper, the two opposing forces on the piece of paper are comparable and the particles that cause them should be just as easily detected. The fact that it has escaped detection for so long, is suggestive that the graveons, if they exist, need to be operating such that they cannot easily be observed. This treatise will describe the operation of a multidimensional mechanism that shields it from being detected. Furthermore, the graveon does not pull matter together; it pushes matter together.

Immediateness of Gravitational Force

When Newton first calculated the movement of the planets about the sun, he assumed that gravity always acted in the direction along the line adjoining both celestial objects, to get the correct orbits. He then came to the conclusion that if the action of gravity was along the line connecting both objects, the action must be immediate, that is, gravity must travel through space with no delay. He reasoned that if the force of gravity were delayed, then the direction of forces between a planet and the sun, for example, would be based on the relative position of the two objects before its present position. This would introduce a slight torque tending to speedup orbital motion. Eventually, all objects would spin away from each other. That, of course, does not happen.

Various tests have been run that prove that gravity is acting along the adjoining line between objects. One of the more convincing of these tests, involves measuring the gravitational forces that the sun and moon exert on Earth during a solar eclipse. At alignment, the gravitational force is largest, because both the sun and the moon are pulling in the same direction. However, the maximum eclipse occurs after the maximum gravitational force, signifying that the gravitational peak occurs when the eclipse happens, not when it is seen, minutes later. In technical lingo, gravity has no aberration, but sunlight has 8.3 minutes of aberration. However, though gravity acts along the line adjoining the sun and Earth, that fact is not sufficient to prove the force is immediate.

Although Newton did not know the mechanism for gravity, he assumed the most conventional paradigm for gravity; that their pulling force flows out of each object reaching the other object with potentially some delay. This led to the conclusion that a delay in the transmission of this force would produce a torque increasing the orbital speed. This doesn't happen. Consequently, he assumed the delay was zero. For a different paradigm (e.g. LeSage mechanism to be discussed in Chapter 4), a torque in the other direction could occur, slowing the orbital speed. It depends on whether gravity is a pulling or pushing mechanism. Still, other paradigms (e.g. a combination of the previous two) might not produce a torque in either direction. Then, the force would act along their adjoining-line connection, in spite of having a transmission delay.

While the Newtonian model predicted most of the planetary motion about the sun with no detectable error, one notable exception was Mercury, the closest planet to the sun, with its perihelion advance. General relativity, a refinement to the Newtonian model correctly predicted Mercury's perihelion advance. Just like the Newtonian model, general relativity assumes that the forces between objects act along the straight-line connection between the objects. However, because he wanted to be consistent with special relativity that required no travel faster than lightspeed, he proposed a mechanism for the force in general relativity, which would travel at the speed of light. This mechanism defines gravity as not being an ordinary force but a warping of space. This warping action is so devised as to allow it to travel at the speed of light and still act without an offset. Both the Newtonian and general relativity models make assumptions about the speed of gravity, which are not requisites to their respective theories. An excellent discussion, positing that the speed of gravity is much faster than lightspeed is made by Tom Van Flandern at Meta Research.

The gyroverse theory predicts the force of gravity is along the adjoining line, and acts many orders of magnitude faster than light.

QUANTUM ENIGMAS

ELECTROMAGNETIC FORCE

Charged objects attract each other if the sign of their charge is opposite and repel each other if the sign of their charge is the same. A common manifestation of the attraction mechanism, "electromagnetic force," is static electricity, when your clothes cling to your body, and require the use of special washing products to prevent it. The repulsion mechanism is depicted in slapstick comedy, when someone touches an open electric outlet and his hair stands up. While this will not work because the voltage is not high enough, a Van de Graaff generator that can generate 20,000 volts is often used for this literal hair-raising demonstration. As previously explained a mechanism for repulsion is straightforward. Photons (light particles) jump back and forth between the same polarity charged objects. When the photon leaves an object, it recoils, moving away from the other object where the photon is headed. When the other object captures it, the impact causes this object to move further away from the first. It is similar to two people on skates repeatedly tossing a basketball back and forth to each other. Photons jumping back and forth, force the objects to move away from each other, effecting repulsion.

Clearly, the paradigm that is so intuitive for repulsion does not work for attraction, since each photon would have to leave from the rear and strike the rear of the opposite charged particle to drive the two objects toward each other, like a boomerang. There is no evidence that photons travel in any way other than in straight lines. Therefore, just as with graveons, physicists have come up with a similar paradigm to explain attraction and repulsion, in which the photon is not causing the movement directly, but merely, mediating it by exchanging photons.

This phrase reminds me of a visit to a skin doctor about a chronic rash diagnosed as eczema. He prescribed a cream to be applied twice each day to the infected area. After about two weeks of using the prescription, the rash did not get any better, and I made a second visit, and the doctor prescribed a salve instead. When I went home, I looked up the word eczema in a medical encyclopedia, and to my surprise it said that eczema was not a disease, but was a set of symptoms. Instead of having to diagnose a rash, the cause of which he could not identify, the medical profession has given this broad collection of symptoms a disease-sounding name, creating the illusion of causal identification.

Likewise, current physics cannot explain the electromagnet attraction/repulsion mechanism, so instead invent a phrase, "mediated by exchanging photons," conjuring an image that science understands it, when the opposite is true. Each particle must interpret these messages delivered by photons and act accordingly, using some unknown mechanism. Photons do not possess intelligence, so they cannot convey this message. Even if they could bring the message, what would cause objects to move? It cannot even be blamed on space warping, since none is associated with electromagnetism. This implied mechanism for attraction and repulsion is illogical and gets a failing grade on the Occam's razor test.

Only in higher dimensional spaces is it possible to arrange things such that photons can travel in a straight line from the back of A to the back of B, while A and B face each other, causing attraction. The physical mechanism that can be both, attractive or repulsive, will be described in this treatise.

PARTICLE/WAVE DUALITY I

When light is passed through a narrow-slit impinging on a photographic plate, instead of seeing a sharply defined image of the slit shape, light appears most intense just behind the slit gradually diminishing in intensity as the image moves off-center. This phenomenon is called "diffraction," or the bending of light around a corner. If this slit is blocked and a second adjacent slit is exposed, it too will experience the same image as the first slit, but offset by the separation of the slits. If both slits are opened simultaneously, instead of seeing the superposition of both images, an "interference pattern" will result. Since light is a wave, this is to be expected, because other waves, like water or sound, also experience this interference effect.

Nevertheless, if the intensity of the light is decreased to the point where photons leave the source one at a time, the same pattern results. This is amazing and completely unexpected and is only possible if each photon goes through both slits simultaneously, as a wave interfering with itself. Moreover, if the plates are sensitive enough, each photon makes a single mark on the photographic plate indicating the photons are particles also and must have gone through only one slit. Within the same test, the evidence suggests that each particle is one item going through both slits simultaneously. Equivalent experiments and similar results are observed if electrons instead of photons are used. Thus, each photon and each electron has both wavelike and particle-like properties. In fact, all atomic particles would exhibit these same dual properties. Within the contemporary assumed structure of space, it is not understandable how both characteristics could be valid simultaneously, yet it is an accepted explanation. The gyroverse's structure of space, will show how a single atomic particle can go through two separated slits simultaneously, and interfere with itself.

PARTICLE/WAVE DUALITY II

Another, more common explanation of the wave function that mitigates the previous paradox is that the probability function is doing the waving and interfering. This interpretation of the wave equation, which has its own inconsistencies, was proposed by Max Born and was a pillar of a set of ideas, referred to as the "Copenhagen interpretation." This name was derived from its most ardent supporter, Niels Bohr, who headed the Institute for Theoretical Physics at Copenhagen University where he was Professor of Theoretical Physics. It is more intuitive that something can be a particle and have a probability distribution that is a wave function, than it actually is a particle and a wave function simultaneously. In this interpretation, the photon is strictly a particle, but its location is determined by a probability distribution. The peculiar nature of this interpretation is that the state of the particle remains a probability distribution until the particle is observed. It is not that the state is decided, with the result not being known until observation is made, but instead that the final state does not occur until the experiment is observed. It is not just that the theory predicts this, but innumerable experiments have verified that this interpretation consistently gives the expected results.

If only atomic particles have this peculiar characteristic, the strangeness is tolerable, because it is not an everyday experience. However, Erwin Schrödinger, the great physicist who formulated the wave function that initiated this conflict, believed that this interpretation was wrong, and that some vital understanding of quantum physics was missing. To illustrate the absurdity of this interpretation, he proposed a clever experiment combining a common experience with particle physics. This thought-experiment, known as "Schrödinger's cat," is a diabolical plan to initiate the killing of a cat in a box using radioactive decaying particle that breaks open a cyanide flask. Until someone opens the box the particle neither decays nor doesn't decay, the cyanide flask neither breaks nor doesn't break, rendering the cat in limbo, both dead and alive simultaneously. When the box is opened and observed, the decay either does or does not occur, and the cat meets its final fate retroactively—huh?

While Erwin Schrödinger proposed this exercise somewhat tongue-in-cheek, it did highlight some absurdities that this interpretation leads to; how probabilities, which are inanimate abstract notions, can go through slits and interfere with themselves. Nevertheless, this is still the state of affairs over 75 years later. The gyroverse theory has a much simpler explanation to clear this up.

STRING THEORY

String theory is the latest attempt to unify relativity, quantum theory, and the four forces of nature. String theory starts with the premise that all atomic particles are the result of strings vibrating at each particle's characteristic frequency. While some hope that it will lead to unification, it has its own set of problems, created by a dilemma. The mathematics predicts that some particles in the universe have a negative probability of existence, unless the universe has 10 dimensions. In the words of Freeman Dyson,

"Nothing can happen less often than never." Consequently, the physics community chose the lesser evil, and assumed that the universe was composed of 10 dimensions, which also had some other benefits--not very reassuring. More recent developments modified it to ten or eleven space-time dimensions where only four of them are infinite, as we now know them. The remaining six or seven are curled up so small and tight (compactified) that they are hidden from observation--not even atomic particles can fit through them. The inclusion of extra dimensions provided a mathematical device for attaining unification of the four forces of nature, the theories of quantum mechanics, and relativity. Because of the lack of physical evidence for their existence, they are assumed small enough to keep them hidden. In other words, they provide little other function than unification of the physics involved. It is unreasonable to expect that the extra dimensions, if they do exist, would also not have some physical significance in explaining nature.

Beyond relativity and quantum theories, the widely accepted inflation theory is needed to explain the origin of the universe. While inflation theory is consistent with relativity and quantum physics, they do not predict it. Any unification of physics would not be complete unless inflation theory is also consolidated within string theory.

While the geometric description of string theory would not shed any light on the anomalies presented here, the aspect of it that is important is the notion that extra dimensions might simplify unifying physics. First, the most questionable aspect of the gyroverse model, the extra dimensions, becomes a bit more palatable, since many leading physicists already accept the idea that the universe has many more dimensions. Second, the extra dimensions that improve the mathematical description of nature, if modified, might also be applicable to this theory. Several different formulations of string theory already exist. Perhaps one variation could be retrofitted to apply to the gyroverse model. A third aspect of string theory that is interesting is that each atomic particle type is not envisioned to be a solid object, but a vibration of a subatomic string at a particular frequency. This is consistent with the gyroverse notion that matter has no intrinsic mass, but includes a mechanism for obtaining mass like qualities from a tight bundle of waves. As of yet no attempt was made to apply string concepts to the gyroverse model. The major physical difference between string theory and the gyroverse theory is that in string theory the three recognizable dimensions are infinite while the extra dimensions are only Planck-length, where as all of the dimensions in the gyroverse are of atomic proportions, and differ in kind and number.

QUANTUM ENTANGLEMENT

Of all the enigmas, this stands out the most. Two electron or photon sisters that have a common entangled origin, continue to have an acute awareness of each other, even when separated by considerable distances. Change to one of the particles is reflected in the other. This has been verified with experiments run by Nicolas Gisin, at Geneva University through tests where the polarization of an entangled photon is changed because of changes to the polarization of its sister photon seven miles away. These tests were enhancements to the earlier acclaimed tests run by Alain Aspect and his team at the University of Paris. It defies all semblances of logic that a photon can communicate its polarization to its entangled sister seven miles away. Even more amazingly, this communication was measured to happen at least 10 thousand times faster than the speed of light. Because quantum theory is invariant to spatial separation, it predicts that the communication is instantaneous, even if the photons were on opposite sides of the universe. The only explanation that seems reasonable is that although these sister photons appear to be seven miles from each other, they are in reality very much closer to each other than the trillions upon trillions of photons that appear closer. While this explanation seems impossible, it is possible in a space composed of many more than three dimensions. The gyroverse will describe this mechanism. In contrast to current predictions, entanglement breaks down at some distance, so that entangled particles could not exist on opposite ends of the universe. In the Quantum Theory Overview chapter it is described how a variation of Gisin's experiment can be used to estimate the distance limitation of entanglement.

COSMOLOGY ENIGMAS

UNIVERSE INFLATIONARY MODEL

Hubble was first to establish that the universe is expanding uniformly. More precisely, all matter in the universe is receding from all other matter at a rate proportional to their distance apart from each other. In equation form, for a distant galaxy, $v=Hd$, where H is the Hubble constant, d is the distance from Earth to the galaxy, and v is the velocity that the galaxy is moving away from Earth. On the surface, this expansion is consistent with relativity theory, which predicts that the universe is either contracting or expanding. This uniform expansion causes two problems that cosmologists have labeled the "flatness problem and the horizon problem."

The flatness problem comes about because calculations suggest that for the universe to have evolved allowing matter to have clumped together into stars and galaxies, the force of expansion at the big bang had to be balanced almost perfectly with the gravitational attraction during its initial expansion. This would have been a very unlikely accident of nature. If the expansion force were too small, the gravitational forces between celestial matter would have halted the expansion and started it to contract, causing the collapse of the universe long ago. On the other hand, if it were too large, matter would have raced away from each other at such a high speed that it would never have packed together to form the stars and galaxies. Cosmologists have determined that this initial expansion had to be balanced to less than one part in a trillion to be as flat as it is today. To present an analogous situation, consider a rocket shot up in space. If it does not travel at escape velocity, it falls back to Earth in short order. However, if it travels above escape velocity, it gets far beyond the gravitational reach of Earth, continuing to travel at its ultimate escape speed forever. If it travels almost exactly at the escape velocity, it continues moving away from Earth almost indefinitely, going slower and slower, approaching, but never reaching a zero speed. The universe began its expansion about 14 billion years ago. For it still to be anywhere near flat, it had to have originally expanded extremely close to the escape speed to within an unlikely high precision. Thus, cosmologists determined that the original explanation of the big bang was improbable.

The horizon problem was recognized when the cosmic background radiation, abbreviated as CMB, measured in all directions of space was found almost identical. This cosmic microwave background radiation is a remnant of the early years of the universe, when it was 300 thousand years old. After then, radiation could travel unencumbered in the universe. For the radiation in all directions to be so uniform today shows that matter at the outer reaches of the universe in all directions must have been homogeneous earlier. This would have been highly unlikely, with the original theory, because the regions where the radiation came from could never have been in causal contact with each other. For example, the universe has been observed at distances of 13.5 billion light-years in each direction, indicating that light would have taken at least 27 billion years to travel from one end to the other. These places could not have been in contact since the universe is thought to be between 13 and 15 billion years old. Yet, the radiation coming from all directions being so uniform in intensity showed that they must have been in contact. This is further evidence that the original big bang theory is problematic.

In 1981, Alan Guth, with further important contributions by Andrei Linde, Paul Steinhardt, and Andy Albrecht, developed the inflation model. It was proposed that the universe went through a very rapid expansion initially during its origin. This theory was important because it provided a potential explanation and resolution of the flatness and horizon problems. To its credit, it capitalizes on some well-known quantum physics principles, quantum (vacuum) fluctuation, where matter appears temporarily, from nothing, allowed by the uncertainty principle. However, it is overreaching to apply something from quantum theory to the whole universe. The universe might have started small enough to rationalize it, but it became immense quickly. In this theory, gravity reversed and became repulsive, and the entire universe expanded uniformly with matter getting further apart at a colossal rate. Many variations of this theory have been proposed, but they all allege that the universe has doubled in size over 100 times in about 10^{-33}

seconds, after the big bang. During this expansion, matter was created out of nothing by the vacuum fluctuation and contains zero energy overall.

The expansion is sometimes explained by the metaphor of an inflating confetti-studded balloon. As the balloon inflates, the space between the confetti separates uniformly. In addition, no location on the balloon is preferred. At the end of the inflation period, the size of the universe grew many times larger than the visible part of the universe is today. Distances between matter uniformly increased, without the matter moving in the conventional sense. Another analogy that is also used is raison bread. As the dough rises, the raisons move uniformly apart, separating in all directions. Actually, this metaphor is more fitting for current theory, and the balloon metaphor is more appropriate for the gyroverse paradigm.

While inflation theory, widely accepted today, resolved the flatness and horizon problems, another theory had to be introduced that is neither part of quantum mechanics nor relativity. In a three-dimensional Euclidean space model, there is no mechanism for this expansion to take place. It is not that it conflicts with these theories; it is just that it is not predicted by either of them. If it happened, it has to be explained and fit with the other two theories. The gyroverse model without resorting to inflation, also explains the initial period of the universe. In fact, all physics is explained by just one theory.

UNIVERSE EXPANSION MODEL

Growth of the universe is thought to have evolved through two distinct periods. The first period is inflation, just described, that completed in an infinitesimal fraction of a second after the start of the universe. It grew from smaller than an atom to something much larger than the present-day observable universe. Immediately thereafter, the expansion period picked up, continuing to the present time. Our entire current observable universe originated from only a tiny portion of the inflated universe. Thus, our entire observable universe was born from only a tiny sliver of a subatomic seed of matter. The breadth of the extended universe is believed to be more than a billion times greater than the 30 billion light-year span of the observable universe. However, the expansion that followed the inflation period is problematic.

If it is considered that all matter is moving away from all other matter proportional to their separation, the outside edge of the observable universe is expanding near the speed of light. Extrapolating from this, the extended universe is expanding more than a billion times faster than the speed of light. During the inflation period, it would have been expanding multiple billions of times faster than that. Clearly, this rapid universe expansion is not consistent with special relativity. However, the expansion gets around this limitation by hypothesizing that the expansion of the universe is not ordinary motion, but is expanding space, which presumably can allow relative motion greater than the speed of light, and not violate general relativity.

Often it is argued that relativity has met every threat challenging its veracity. However, it seems that wherever a breach occurs, theories are twisted to skirt apparent contradictions. Inflation and expansion are two such paradigms. Several other instances of this are identified in this book.

SUMMARY

SUMMARIZING ENIGMAS

For all the enigmas to make sense, the real universe must be very different from the way it appears. However, the large body of experimentally verified physics must be consistent with any model. The two most successful theories in modern times are relativity and quantum mechanics. Whenever they have been put to an experimental test in their realm of applicability, which have been often, both were verified to astounding accuracy. Success of that size implies that the formulas are correct or very close to correct. Therefore, any new theory must surely leave these equations intact or at most only slightly modified.

In addition, the model must explain why it was successfully hidden from view for so long. This question will be answered at two levels. On the first level, the cornerstone of the theory has the universe configured in a way that hides its true nature. It has twelve dimensions, with all matter blazing away at the speed of light. At that speed, our universe is trapped within a three-dimensional Euclidean subspace and nothing can escape easily. Obviously, that is not the complete answer. On another level, all the previous enigmas provide prima-facie evidence that the extra dimensions are not perfectly hidden. Until now, these clues have not been recognized. It seems that a model has not been devised where these strange behaviors would appear natural.

When these mysteries are reviewed as a group, the nature of the universe begins to emerge. The energy-matter equivalence implies that the energy of matter is due entirely to its moving at the speed of light and not that matter is energy itself. From the equivalence of the mass of inertia, gravity, and energy, it can be deduced that there is one underlying cause for all three. The only way that two atomic particles separated by miles, can remain in contact is through a short cut in space. There is no other reasonable explanation. The fact that the communication between the entangled photons is so much faster than light, reinforces this hypothesis. String theory and Kaluza-Klein theory provide additional evidence that a space having more than three-dimension is very plausible. Therefore, just by examining these oddities of nature, one can make a good guess as to the nature of the real physical structure of the universe. A next step is to take this as a starting point and build a cohesive, self-consistent construct of the universe that can tie all this together and explain all the enigmas. That is accomplished in this treatise.

PREDICTIONS - POSTDICTIONS

The term prediction, in physics, means to foretell the results of physics experiment that are not presently known. For example, Einstein, to show the efficacy of general relativity, predicted the amount that light rays passing by the sun would be bent due to the sun's gravity. The test could not be run until the next solar eclipse. When the experiment was run, his prediction was confirmed. Before physicists take a theory seriously, they want to see some confirmed predictions. If the new theory cannot make any new predictions, then why is it better than the old theory? Postdiction, made-up physics' jargon, predicts after-the-fact events, an oxymoron. Clearly, postdictions are not as valuable as predictions, because the proponent is suspect for having artificially twisted the theory just to match known results. Though Einstein had several recognized major contributions in physics to his credit before he published general relativity, it took considerable cajoling on his part to get the testing done. At one point, he was even considering using some of his own money. The proponents of string theory, many of the most renowned physicists of our time, have made some predictions that require new super particle accelerators to verify their theory. It will be some time before those tests can be run. As physics gets more complex, the ante bar for verification is raised a little higher.

The evidence offered here comes under the postdiction definition. Yes, it is different because it does not attempt to explain one or two of the unusual quirks of nature, but over a dozen of them. The theory is radical, but it takes radical thinking to make sense out of so much of today's physics.

CHAPTER 3 - THE MODEL BASICS

INTRODUCTION

In the previous chapter, the case was made that if Euclidean space is examined, something is amiss with the current model of the universe. In fact, if the model is examined critically, hints as to the real nature of the universe emerge. Evidence points to our universe being embedded in a space of many more than three dimensions. Unfortunately, all our thinking about higher dimensional space is tainted by our intuitive experience with three-dimensional space. In the midst of developing this theory, it became apparent that some fundamental parameters of nature only make sense in three dimensions. Mass, inertia, momentum, and even energy can only be defined in three dimensions. For a higher dimensional space, it is not clear if alternative notions exist. Our perception of time would take on a completely different character in a higher dimensional space, since time is related to inertia. Because the idea of additional space dimensions is a very significant departure from the prevailing perception of the universe, additional dimensions will be explained building on familiar concepts. These ideas will be developed in three-dimensional spaces, where visualization is readily attainable and extended to higher-dimensional space, where the ideas are needed but visualization is difficult, making the problem more tractable.

BALL OF STRING METAPHOR

Start with a ball of string about the size of a baseball and one-dimensional creatures that live on it. For them, traveling from one end of the string to the other end would be a large distance. For example, if the string were about a hundredth of an inch in diameter, the one-dimensional creature would have to travel over a mile from end to end. In three dimensions, nothing in the ball is further than three inches apart. However, even a human hair is much thicker than what is needed for the living space of a zero-thick, one-dimensional creature. Nevertheless, a string the thickness of a human hair would stretch about twenty miles, and yet the furthest distance apart on the ball would remain three inches. Continuing that train of thought, if the string were one atom thick, the string would be about a 100 trillion miles long from end to end. Again, the maximum distance between any two points in three dimensions would be 3 inches. In fact, a string the thickness of a single atom is still much wider than what is necessary to contain a one-dimensional creature. As the string width approaches zero, the length of the string would approach infinity. Additionally, the ball could shrink to the dimensions of a grape, a pea, a grain of sand, or even an atom, and yet an infinite string without any width could still be wound around it. This is because a string without width can have an infinite number of windings, and still not occupy any volume.

CARD FILE EXAMPLE

Just as an infinite one-dimensional string, can be contained in a small three-dimensional space, even an infinite two-dimensional sheet could be contained in a three-dimensional, small space. This can be illustrated by imagining a three-by-five card file where the cards are made thinner and thinner. If the cards are reduced to zero-width, the number of cards contained in the file could increase to infinity. If the cards were removed and placed side by side, the sheet would have infinite area. The cards could be made smaller, and only more cards would be needed. In addition, the file could be shortened, since zero-width cards take up no space in that dimension. An infinite number of them could be contained in a short file. Using this reasoning, it follows that an infinite space of any dimension can be contained in a small space of a larger dimension. Vast distances in the infinite space could be kept very small in the higher dimension space.

Hyper-Box Simile

Consider another example, a four-dimensional hyper-box, closer to the model, but more difficult to visualize. In three dimensions, a four-dimensional hyper-box becomes an ordinary box. However, the entire three-dimensional universe would fit into this four-dimensional hyper-box, even one of atomic proportions, about an angstrom on its edge. This can be visualized with some help from the card file example, except that an additional dimension is added so that boxes instead of cards are being filed. At each of the infinite number of points along the fourth dimension, another empty box is found with just enough room to place a three-dimensional atom in it. Since atoms take up no room in the fourth dimension, they can be placed arbitrarily close to each other in that dimension.

Figure 3.1 – Four-Dimensional Hyper-Box

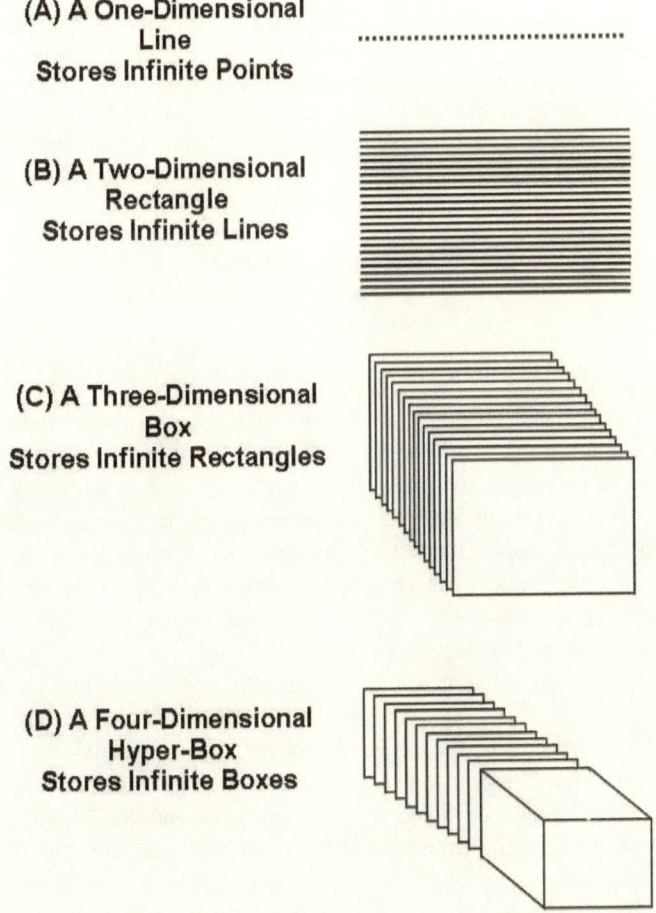

(A) A One-Dimensional Line
Stores Infinite Points

(B) A Two-Dimensional Rectangle
Stores Infinite Lines

(C) A Three-Dimensional Box
Stores Infinite Rectangles

(D) A Four-Dimensional Hyper-Box
Stores Infinite Boxes

Figure_3.1 - This figure illustrates how a tiny four-dimensional hyper-box can have enough inside space to hold every atom in the universe. The top illustration (A) shows a one-dimensional line that can contain an infinite number of zero-dimensional points. Illustration (B) portrays a two-dimensional rectangular sheet that can contain an infinite number of lines. The third illustration (C) depicts a three-dimensional box that can contain an infinite number of rectangular sheets. Keep in mind all of the points, lines, and sheets have finite width in order to be seen. In reality, these objects would have no width and they could be packed at an infinite density.

By inference, a finite larger dimensional container could hold an infinite number of lower dimensional objects. Finally, the last crude illustration (D) depicts a four-dimensional hyper-box that could store an infinite number of three-dimensional boxes. Each three-dimensional box has to be only big enough to fit one atom inside it. Thus, the box size need only be an angstrom on its edge, about the size of an atom. Since the boxes have no thickness in the fourth dimension, it can be packed at infinite density in that dimension. It would take about 10^{80} three-dimensional boxes to hold all the atoms in the universe, a huge number, but still much less than infinity of boxes that the hyper-box could contain.

This is analogous to the card file example, where the cards can be placed arbitrarily close to each other in the third dimension. It would take 10^{80} (a 1 followed by 80 zeros) of points to place all the known universe atoms into it. This is a large number of points, but much less than the infinity of points contained in this fourth dimension. If one wanted to include the empty space in our universe, 10^{110} of points are needed, which is still much less than infinity. Although all distance measurements in this four-dimensional hyper-box are small, the volume of the box is enormous. The idea can be further explained with the help of figure_3.1.

However, while the universe, as we know it would fit into a small four-dimensional space, it would not fit in it neatly. It would have to be folded or diced up to make it fit, similar to what was done in the box file example. This would not have the look of the universe. As the number of dimensions increases, the size of the available space becomes infinitely larger, allowing the universe to fit in it much neater. If the number of dimensions is increased sufficiently, the universe could be "rolled-up" in it, and retain its Euclidean appearance of our familiar three-dimensional subset.

EDWIN ABBOTT'S FLATLANDERS

Perhaps the hardest thing to fathom is where these extra dimensions might be. One cannot point and say; "It is in that direction." While giving a completely satisfactory answer is impossible, the classic 1884 book by Edwin Abbott, an English cleric, called <u>Flatland</u>, came closest. This book about Flatlanders, a race of people, is an allegorical satire on Victorian England's societal classes. It is illustrated through geometry of two dimensions where women are represented by straight lines, and men by closed figures, ranking based on the number of sides they possess. Triangles are soldiers; circles, actually many-sided polygons, are the priestly class, and every other status lies somewhere between.

An interesting idea in this book is how the narrator A. Square tries to reveal his experience with a three-dimensional sphere that he met, to other Flatlanders. He can only point in the four surface directions, and not up or down, making it very difficult to describe the third dimension that no one has seen. When a sphere moves up and down through flat space, a circle of a changing radius appears. This sequence of events is not possible to explain in two-dimensional Flatland and is a strong indication that something is amiss. We recognize the top of a Flatlander, but to a Flatlander the notion of top is foreign, because Flatlanders have never seen their top. Similarly, we are three-dimensional and have the same difficulty in experiencing or describing the fourth or higher dimensions that may exist, because we cannot describe where they are. The following figure shows a Flatlander soldier.

Figure 3.2 – Flatlander Soldier Class

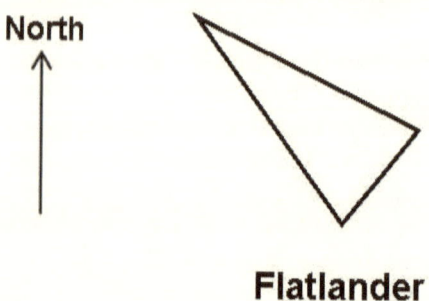

Figure_3.2 - This figure illustrates a soldier from Edwin Abbott's geometric parody on Victorian England societal classes. This Flatlander soldier is now pointing northwest but can turn to point in any other direction. However, he cannot point in or out of the paper. He has never seen his topside or any other Flatlander's topside. Just as a two-dimensional race cannot see or easily understand the notion of three dimensions, we cannot easily grasp four dimensions.

As hard as it is for Flatlanders to experience three dimensions, doing it is possible. Imagine a plateau of Flatlanders and a hillside of Flatlanders separated by a canyon portrayed in the following figure. Flatlanders on the plateau can view the top of Flatlanders on the hillside as they climb the hill from the canyon. The entire top of the hillside Flatlander cannot be seen all at once. As a soldier Flatlander climbs the hill, the plateau resident first sees the front apex of the Flatlander. Subsequently, only a slice of the top is seen during the climb, which grows in size and disappears all at once as the hillside Flatlander moves out of view. The gradual growing is not unique, but the sudden disappearance of an edge cannot occur as Flatlanders view each other on the plateau. The Flatlanders might discern that something is awry in what they are seeing, and could deduce it to be the top of other Flatlanders, illustrated in the following figure.

Figure 3.3 – Different Flatlanders' Terrain

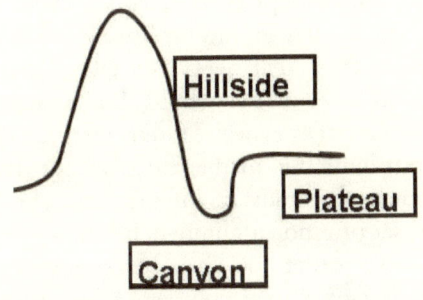

Figure_3.3 - This illustration shows two locations of Flatlanders. One is on the plateau and the other is on the hillside. Flatlanders on the plateau can view Flatlanders on the hillside, but only a single slice at a time.

GYROHELIX PRECURSOR

Just as it possible for Flatlanders to ascertain that they may possess a topside in the third dimension, observing fragments of our extra dimensions is possible, if we are perceptive. As atomic particles are accelerated to high speeds, it appears that the particles get shorter. This shortening is due to fact that a particle in motion is completely visible only in three dimensions slightly offset from our three-dimensional view. This places the particle partially in our fourth motion-dimension, which is out of our view. Since this fourth motion-dimension is hidden, the length of the particle in the direction it is moving appears foreshortened. Placing a far-off object obliquely to one's view can approximate this effect. From far away, it could be hard to notice that one side of the object is closer than the other side, making it appear shorter. This is one of many hints of the existence of other space dimensions.

CARDINALITY DIGRESSION

When one thinks about infinities of different magnitudes, it is tempting to apply the concept of cardinal numbers, developed by George Cantor, which sorts infinity by size. It has been suggested that it be looked at from that perspective. George Cantor was a mathematician born in Saint Petersburg, Russia, in 1845, moving with his family to Germany where he remained for the rest of his life. His major contribution in mathematics was the development of set theory, especially the study of infinite sets. The set of whole numbers and the set of rational numbers belong to this group. The next known higher order of infinity, sometimes called the continuum, is the set of all real numbers, which contain both the rational and irrational numbers.

Unfortunately, the notion of cardinal numbers does not have the necessary granularity. Spaces of any number of dimensions all have the same cardinality. All the examples presented so far have the same cardinal number. Consequently, cardinality does not apply directly to this problem. Nevertheless, before Canter, infinity played a much lesser role in mathematics, and several great mathematicians of his time denigrated his discovery. Though the notion does not apply directly, the idea that not all infinities are the same size is important. While adding more dimensions does not increase the cardinality, the added degrees of freedom allow subspaces of it to include other properties, not otherwise possible, such as being Euclidean.

INTRODUCING THE GYROHELIX

A theory, accepted by many leading physicists, proclaims that the universe contains many more dimensions. This notion, known as string theory or M-theory, ascribes six or seven extra dimensions to the universe. They include four conventional space-time dimensions that are infinite, and six additional ones that are tiny and hidden by their smallness. These extra dimensions are configured in a geometric shape known as a "Calabi-Yau Manifold." The advocates of string theory expect it to unify all physics into one theory. However, a major problem is that analysis in a ten or eleven-dimensional space has the liability of creating extreme analytical and visual complexities.

Since the gyroverse model also assumes many more dimensions, it too could have the mathematic limitations of string theory. However, this quandary was avoided by recognizing the symmetry of three-dimensional space. Rather than attempt to model the three dimensions as one, only one dimension was modeled. It was implicit that the model for the other two dimensions would be the same. This approach enabled each axis representing one dimension to be modeled with just three dimensions. The entire space is visualized by combining the effects on each axis. This is much simpler, both visually and analytically, than attempting to analyze the entire problem in one fell swoop.

Three-dimensional Euclidean space can be defined through the three imaginary orthogonal space directions x, y, and z. To fit this infinite space into a small manifold, the entire space is rolled up, with each axis direction individually rolled up into a three-dimensional helix forming its own space subset of the higher-dimensional space. This is done so that from the outside, the universe is a rolled-up multidimensional helix, but from the inside, it appears Euclidean and infinite.

It has already been alluded to that Euclidean space has four-dimensions. The fourth dimension is the direction where all matter is moving at the speed of light, and is constructed like the other three. This is known because in other parts of the universe, the fourth dimension is one of three.

CHAPTER 4 - THE GYROVERSE

INTRODUCTION

The law of inertia was first proposed by Galileo Galilei. His characterization of the law was a qualitative description of inertia. Sir Isaac Newton took this qualitative description as his first law of motion. To this he added the second law of motion, F=ma, which quantified the first law. The first law of inertia is "A body at rest tends to remain at rest. A body in motion tends to remain in motion. Either of these conditions persists unless a force is applied to change it." At the same time, Newton also proposed the law of gravity, which can be expressed by the relationship $F - G m M /r^2$.

One interesting sidelight to this story is his well-known response after repeatedly being asked for the root cause of gravity, not merely its mathematical formula. In an appendage to the third addition of Principia, he admitted not understanding it and answered his critics with "hypotheses non fingo," which in latin means "I frame no hypotheses." It is significant that both he and the reviewers expected to have a causal explanation alongside the equations. We still don't have an adequate explanation, but no one cares anymore.

Additionally, dating from Newton, it has never been resolved why gravitational mass and inertial mass are the same. Newton conducted experiments to measure the difference but could not find any. Ernst Mach thought that all of matter's gravity in the universe collectively caused inertia, but could not justify his belief. Einstein, in general relativity, just postulated their equivalence as a first principle of nature, which unfortunately quelled further investigation of this enigma. He initially sited Mach's work, but later lost interest when general relativity solutions were discovered without the need of any matter in the universe. While the gyroverse does not assume this equivalence, a close association between these two phenomena will be explained.

Before that equivalence can be addressed, the heart of this theory, the gyroverse model, has to be understood. Standing in an open field, peering at a night sky, one has to sense the vastness of the universe. Realizing that what is seen is only an insignificant part of the whole universe, which is 14 billion light-years in every direction, makes it hard to grasp. What will be learned is that the entire universe is packaged physically into something of atomic proportions. From end to end in every direction, its width is approximately an angstrom. It is so small that from the outside, inches away, it would be too small to be seen with a microscope. It could be straddled between your legs, given a place to stand. Curiously, the abundance of empty space on the inside makes it hard to explain why nothing gets lost. Finally, from the outside, it is at least 28 orders of magnitude less substantial than it is in the inside, bordering on being nothing. While all this can be explained intellectually, it cannot be understood emotionally. So, leave emotions at the door to begin understanding the gyroverse model. This theory is referred to as a model because while it captures the essence of the structure of the universe, I suspect it is only an approximation to its actual construction.

GYROVERSE MODEL

It seems implausible that the universe can be rolled-up to the size of an atom, still appearing Euclidean and infinite on the inside. Higher dimensional spaces are usually studied from their own multidimensional perspective, and not thought of as a repository for much larger, lower dimensional spaces. However, this is exactly what the gyroverse is; a microscopic twelve-dimensional manifold containing our almost infinite three-dimensional space. This treatise will explain how this is possible.

Our Euclidean space consists of the three recognizable dimensions referred to as "position-space." They are also referenced by the three orthogonal axes, x, y, and z. Each inertial frame is moving in a fourth direction called the w-axis or motion-direction. An observer on their own inertial frame, which it views as the rest frame is not aware of this motion or of any other motion in their motion-direction. Other inertial frames moving relative to the rest frame are also moving at the speed of light, but slightly offset from each other, so that each can only observe their relative motion to others. For ordinary motion, this difference of position-space is too small to be noticed.

Figure_4.1 represents one axis of Euclidean space. It is coiled as a helix. Each revolution of the axis (circumference) on the hyper-cylinder is in the order of magnitude of 10^{-9} centimeters. The advance along the helix per revolution, its pitch, is about 10^{-45} centimeters. Each of the other axes appears the same as this one but is in a separate set of three different dimensions of space. Because of the rotational and translation symmetry of space, it does not matter where the origin is assumed and in which direction the axes are drawn as long as they are mutually orthogonal to each other. Three copies of Figure_4.1 make up position-space, the three spatial dimensions, x, y, and z. Beyond the normal three directions, all matter is moving in a fourth dimension, its motion-direction, at the speed of light. It too is rolled up, so that the motion revolves along the fourth dimensional motion-direction helix. The angular velocity, called roll, of all matter on the motion-direction helix is the speed of light divided by the radius of the helix. This fourth direction can be modeled similarly to the others, making the entire space model twelve dimensions. In their own motion-direction, objects move at the speed of light, but in position-space only their relative speed, which is much slower, is observed. At a helix-pitch of 10^{-45} centimeters, and assuming all matter in the universe has been traveling at the speed of light for about 14 billion years, matter would have advanced along the helix about two tenths of an angstrom.

Each hyper-cylinder is really a small section of a hyper-torus, which is not drawn to scale. The horizontal magnification of the helix, as drawn, is roughly 36 orders of magnitude greater than the vertical magnification. Through the gravity discussion, later in this chapter, it will become apparent why the hyper-torus is a more logical supposition for the shape of the universe than just a long hyper-cylinder. In short, this is needed to insure graveons circle back and are not lost forever. Similar to string theory, instead of the helixes being constructed in Euclidean spaces, perhaps the extra dimensions take on a torus shape. In that way, the gyro will be constrained to the surface of the hyper torus, not needing an undefined constraining force. Also, the four-way hyper-torus will also require 12 dimensions These extra dimensions, like in string theory would be small enough to elude detection. Also, the fact they're in the universe's super space makes even more difficult to detect. A small section of the hyper-torus can be approximated as a cylinder.

Figure 4.1 - Representation of One Axis

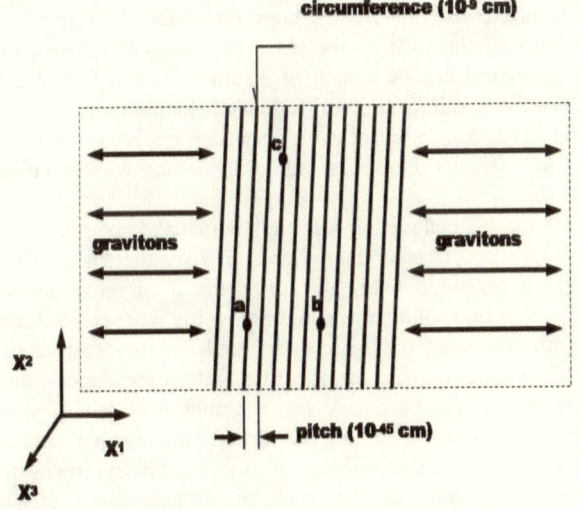

Figure_4.1 - The figure is a side view of an object with a cylindrical shape. Each positional axis is contained in a separate three-dimensional subset of the twelve-dimensional space, where each axis is wound as a helix. This diagram is not to scale; for better visualization, the horizontal direction is greatly exaggerated for clarity, magnified 36 orders of magnitude more than the vertical. In reality, the windings are tight up against each other. The first axis, the only one shown, is x_1, x_2, x_3; the second is y_1, y_2, y_3; and the third is z_1, z_2, and z_3. Dimensions w_1, w_2, and w_3 represent the motion axis, which complete the twelve-dimensional model space. All motion in the same inertial plane is along the w-axis at the speed of light, but is stationary in the three positional axes. Matter is circulating the helix at the speed of light exhibiting immense gyroscopic action, fittingly called the gyrohelix. The helixes in position space are sometimes referred to as position-helixes or just helixes. The four double arrow lines on each side of the windings represent the direction of the gravitational force and the graveons that cause it. Because the helixes are pitched slightly forward, only a small component of the gravitational force is acting in the axis direction, about 10^{-36} times the gravitational force in the horizontal direction. Each axis is wound in its own unique set of three axes. When space is viewed from the x, y, and z dimensions, all space appears Euclidean, because the windings are contained in separate space dimensions, orthogonal to each other.

It is not clear why matter hugs the helix. A magnetic field slightly off perpendicular to the plane of rotation has some of the correct characteristics. A stream of electrons would follow a helix path. a proton stream would also follow a helix path, opposite to the electrons but making a bigger circle because the force is the same but the proton is heavier. However, a neutron stream would continue in a straight line.

Since we are really talking about the behavior of matter in 12-dimensional space, outside the universe, perhaps there is a field there similar to a magnetic field, but proportional to mass instead of charge. In that case, all matter would follow the same size helices.

Strictly speaking, the helix only dwells on the surface of the cylinder, so it could be argued that only two dimensions are sufficient to describe each axis. For example, four sets of a closed circular dimensions with an Euclidean dimensions could be configured to replace the four sets of three Euclidean dimensions. It would have the advantage of not needing an explanation of what is constraining matter to hugging the gyro-helix. In addition, it makes the mapping to the four Euclidean dimensions more intuitive.

Figure_4.2 represents three unwound copies of the helix axes in Figure_4.1, one for each of the x, y, and z axes. This shows a chunk of three-dimensional Euclidean space. Each box in the array of boxes represents one complete revolution along the helix of each axis. Several points, a, b, and c are mappings of the same points in Figures 4.1 and 4.2. As already mentioned, Figure_4.2 shows the three position-space dimensions. The fourth dimension can be envisioned by imagining this entire figure is moving uniformly in the w-direction, as shown, at the speed of light. In a four-dimensional representation of space, all the axes are straight lines. In twelve-dimensional space, the axes are actually helixes. The w-axis motion-direction helix, in which matter is moving, is referred to as the gyrohelix, because the circulating matter exhibits immense gyroscopic characteristics. A one-dimensional sub-space of a helix is a straight line. Another way of saying this is that a helix is intrinsically a straight line. A four-way 12-dimensional hyper-helix is intrinsically four orthogonal straight lines. When referring to a helix in position space, it is referred to as a position-helix or just helix.

Points a and b on both figures represent two points, sister-locations, that are at the same relative location of successive windings of the helix. Point c, a random point, not a sister to the other points, is on the order of 10^{-9} cm from points a and b. While 10^{-9} cm is small, 10^{-45} cm is much smaller. To put this size difference in proper perspective, if 10^{-45} cm represented the width of an atom, 10^{-9} cm, which is approximately one tenth the width of an atom, would represent the radius of the observable universe, 14 billion light-years. If electrons were situated at points a and b, they would have an immense attraction for each other, even if they were far apart in three-dimensional space. This is the situation for entangled electrons. They are far from each other in three-dimensional space, but very close in full space.

Figure 4.2 - Three-Dimensional Space Cells

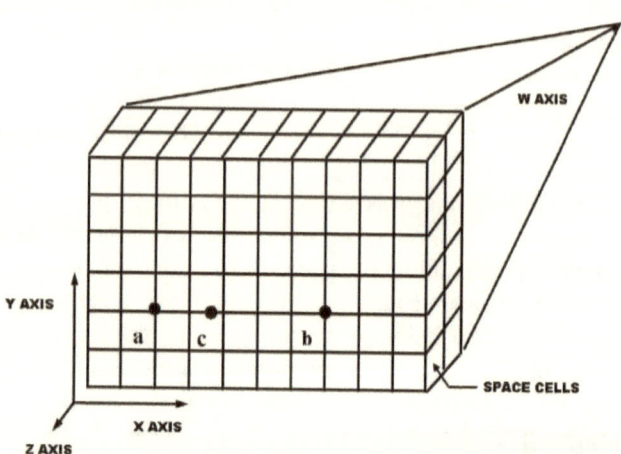

Figure_4.2 - This represents three-dimensional position-space. Each cube edge of the imaginary grid represents one complete revolution of each axis position-helix. These imaginary cubes are called space cells. The points a and b, on both figures, represent the same set of points, and are at sister-locations, on the order 10^{-45} cm apart. Point c, a random point, not a sister to the other points, is on the order of 10^{-9} cm from points a and b. The entire drawing is moving, as shown, in the w-direction.

Carrying the discussion one step further, the first hyper-cylinder is in space x_1, x_2, x_3; the next hyper-cylinder is in space y_1, y_2, y_3; and the last hyper-cylinder is in space z_1, z_2, z_3. This comprises a nine-dimensional orthogonal subspace with all nine axes perpendicular to each other. Each point in our three-dimensional space maps to the x, y, and z-axes, imaginary coordinates wrapped on the three hyper-cylinders. Alternately, a normalized linear combination of the three axes would represent a line in space oblique to the three axes. The union of all such lines in space generated this way would sweep out the entire position-space of the universe for a particular inertial frame.

The Gyroverse
The Hidden Structure of the Universe

All matter, rulers, and light rays, and so on, travel this curved helix path together, helping hide this wound-nature of the universe. This is similar to objects viewed through fiber optic cable that is tightly wound. Since the light from the viewed object hits your eye straight-on, one is not aware that the fiber cable is wound. Additionally, the x-axis is not wound into any of the y, z, or w dimensions; it is only wound within the x_1, x_2, and x_3 dimension group. If the wound nature could be observed in the x-direction, then it follows that the dimensions it is wound into could also be observed, but that does not happen. The other axes are wound the same way, and the same reasoning applies. From the x, y, and z perspective, all space appears Euclidean, because all the axes are orthogonal to each other. There is no curvature across these dimension groups. For example, the x-axis does not bend into the y-axis, the z-axis, or the w-axis. On the other hand, objects are wound in a direction where the objects have no width. Objects can be rolled-up in those directions without applying any stress to the objects, like a painting canvass being wound and unwound with little apparent damage to the painting. When an object is bent, compression develops on the inside curvature and a tension develops on the outside. The thinner the object, the more it can be bent without breaking. If the painting were rolled-up too tightly, some stress and damage to the painting would occur, unless the thickness of the painted canvass is proportionally reduced. With the universe, the thickness in the direction of bending is zero, allowing the winding to be of atomic proportions, and still produce no stress.

Through this subterfuge, from the outside the universe is a tightly wound multidimensional hyper-helix, but from the inside it looks and feels like an almost infinite Euclidean space. Again, large objects can only be rolled-up in dimensions where objects have no thickness. Otherwise, they would break. This, and other factors to be described, constrains objects to the Euclidean subspace.

Take an example of a long thin stick, in Euclidean space, lying on the x-axis and an eyeball also aligned with the x-axis further down the axis. From its wound-up perspective, it still would be lying on the x-axis, because by construction the x-axis with anything on it are wound with the axis onto the x-axis hyper-cylinder. The converse is also true. If the stick is wound-up with the x-axis, when viewed unwound; it will still be lying on the x-axis, because the wound x-axis and anything lying on it will not bend into the y or z axes. Therefore, the stick will appear straight. The fact that the stick has no thickness in the space x_1, x_2, and x_3 where the x-axis is wound, will allow any amount of bending in these directions without generating any tension or compression. Thus, the stick will not break. Light from the stick to the eyeball will also travel along the x-axis, allowing only the edge of the stick to be seen, and hiding the rest of the stick. This is exactly what would be expected in a Euclidean space with a straight stick.

While matter and photons move in the motion-direction at the speed of light, photons also move in the position-directions at the speed of light. However, when matter captures a photon, it only moves with the matter in the w-direction, and would then be at rest, relative to the matter. Two objects moving relative to each other are each really moving in their own w-direction at the speed of light. Each object sees itself as at rest and the other object as moving at their relative speed difference. Understandably, the space dimensions are separated differently for objects in relative motion. Each of these inertial frames resides in the same four dimensions, but with their axes rotated slightly with respect to each other, causing a different partition of position-space and motion-direction. The fact that objects in different inertial frames partition space differently, suggests that the physical characteristics of the four dimensions are symmetrical, since one object's position-space might contain another object's motion-direction.

Subsequently, it will be shown how this peculiar geometry leads to limiting the relative speed between objects to the speed of light, and creates the illusion of non-simultaneity.

GRAVITY MECHANISM

THE GRAVEON

While we experience gravity in Euclidean space, the gravitational force is really acting across the helix transverse to the axis of rotation. This is portrayed as the horizontal double arrow lines in Figure_4.1. The distance traveled along the helix is approximately 10^{36} ($10^{-9}/10^{-45}$) times the same distance traveled in the longitudinal direction, across the helix. The graveons, which cause gravity, travel transversely to the helix at the speed of light, but effectively travel 10^{36} times faster in Euclidean space. Movement across the helix shortens all distances by this same 10^{36} factor. Consequently, the gravitational force across the helix is greater than the corresponding force in Euclidean space by the identical 10^{36} factor, but with a range limited to subatomic distances.

As previously mentioned, since gravity is really acting across the helix, the gravitational force in this direction is 36 orders of magnitude greater than in Euclidean space, but distances are 36 orders of magnitude shorter. Therefore, the energy, the product of force and distance, is identical from both perspectives. The gyrohelix is a ramp mechanism that trades force strength for range. This process is generically called "mechanical advantage." An everyday experience that illustrates this tradeoff occurs when moving stacked chairs. All six can be stacked and carried at once, or they can be stacked in two piles of three, in three piles of two, or left separate. The smaller the pile, the lighter the load, but the more trips required. This is a form of mechanical advantage, trading carrying weight for walking distance. In all of these alternatives, the weight carried multiplied by the walking distance is the same.

A form of a ramp was used to build the pyramids. The blocks of stones used for the pyramids were too heavy to lift vertically. Instead, a long ramp was built so that the blocks could be pulled up a slight slope. For example, to lift a block up 100 vertical feet, a ramp 1000 feet long might be constructed. Neglecting friction, it only takes one-tenth of the lifting force to pull the block up the ramp, but it must be pulled 10 times further. From a construction point of view, trading force for distance is advantageous. If Figure_4.1 is rotated 90 degrees clockwise, the circular ramping action of the gyrohelix can be more easily recognized. However, here the mechanical advantage is 36 orders of magnitude, a multiplier that is hard to fathom.

GRAVITATIONAL FORCE - LESAGE THEORY

Gravity, acting across the helix (Figure_4.1), implies that the graveon, the mediator of this force, also moves in this longitudinal direction. If a graveon leaves one object and impinges another object, the recoil would push back on the originating object, moving it backward. The object it strikes moves forward, increasing the separation of the two objects. This mechanism would cause repulsion, not attraction. Since gravity causes attraction, this is clearly not the correct paradigm for gravity. Curiously, this mechanism would also not explain how a non-ending source of graveons could emanate from an object without getting replenished, since the other nearby objects do not capture most graveons emitted.

LeSage mechanism was originally developed by Nicolas Fatio de Duillier, Swiss born, who was close to Isaac Newton. It was re-invented in the seventeen hundreds by Mikhail Lomonosov, one of the great Russian scientist. This idea is referenced in <u>Six Easy Pieces</u>, one of Richard Feynman's books. Mikhail Lomonosov established the Moscow State University that was subsequently named after him. Nevertheless, his work was not as well known outside Russia. The notion was that small subatomic corpuscles, which now would be called graveons, uniformly traversed all space. These graveons move in every direction and are slightly absorbed by all matter. The total number, per unit time, captured would be a measure of its gravitational mass. All objects, standing alone, would experience no net force in any direction, since they would receive hits uniformly from all directions. However, the uniform showering of graveons would compress the object together. According to this theory, Earth and all its belongings are held together by the graveons showering Earth. This force is recognized as gravity. When two objects are near each other, each object shields a small percentage of the graveons from hitting the other object. The net force on each object becomes unbalanced in the direction of the neighboring object, where a sparsity of graveons causes attraction between the two objects. Also, the total number of graveons captured by both objects per unit time is decreased showing that their total mass has decreased. Captured graveons are subsequently re-emitted, keeping the total number of graveons in the universe stable. Examination of this mechanism would confirm that this attraction mechanism has the characteristics of gravity in that its force is inversely proportional to the square of the distance between objects. This idea did not catch on, but in 1784 the Swiss physicist George-Louis LeSage, revisited this thesis, expanding and popularizing it. The idea now is referred to as LeSage theory. Such notables as David Hilbert, Herman Minkowski, and more recently, Richard Feynman, have also reevaluated and discarded this mechanism.

Figure 4.3 - The LeSage Mechanism

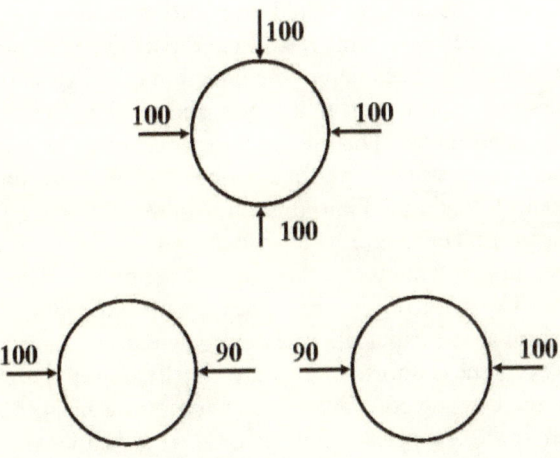

Figure_4.3 – The top schematic is a depiction of a solitary celestial object. It is bombarded in each direction by 100 units of graveons. Including the 100 units in and out of the paper, the total showering is 600 graveon units, which constitutes the gravity holding the object together.

The lower schematic shows two objects close to each other. Each object is also bombarded with 100 graveon units from every direction, except in the direction of the neighboring object, which blocks some of the graveon from hitting the other object. This decreases the number of impinging graveon units from the direction of other object to only 90 graveon units. Each object has an imbalance in the direction of the other object driving them together. The closer the objects are together, the more graveons are blocked, increasing their apparent attraction for each other.

However, as plausible as this paradigm was when originally proposed, it suffered from several shortcomings that caused its dismissal as a viable contender for explaining gravity. First, it could not explain why the graveons were not detected. While a very small percentage of the graveons absorbed by any object are causing the attraction, a five-pound weight has to have a five-pound imbalance of force to weigh it down, which must involve a considerable number of graveons. How could so many graveons escape detection? Second, consider Earth orbiting the sun. As Earth moves around the sun, it would run into graveons, creating an imbalance between the number and energy of graveons impinging on the front and the back end in the direction of the orbital motion. It is analogous to running in the rain and getting wetter in your front than your back. This slight imbalance over several billion years would have slowed the forward motion of Earth enough so that it would have crashed into the sun billions of years ago. As well as causing some aberration which would have added to the orbital instability, not to mention that it also conflicts with general relativity. Consequently, the idea was abandoned from consideration as the explanation for the gravitational force,

However, these shortcomings are not lethal for the gyroverse model, since the graveons are moving across the helix, not in our Euclidean space, so placing a detector in three-dimensional space would not be expected to intercept graveon particles. The second and third concern are also ameliorated in this model. The graveons are traveling at the speed of light in a direction where it only has to travel a distance of about 10^{-8} centimeters to get from one end of the universe to the other end. In three-dimensional space, it is effectively traveling the equivalent of the speed of light to the forth or fifth power. Therefore, the rotational inertia of Earth around the sun is so slow by comparison, that even in billions of years, the inconsequential imbalance of force, would have an insignificant impact on the orbiting speed. Hence, the several major difficulties with the LeSage paradigm are finessed in gyroverse model.

Because of the gyroverse's four Euclidean dimensions, the accepted gravitational mechanism would lead to the force being inversely proportional to distance cubed. This, of course, would be wrong. The LeSage mechanism does not have this shortcoming, and instead correctly predicts the force being inversely proportional to distance squared. In fact, any number of space dimensions would not affect the gravitational formula's distance dependence. This is further evidence that the LeSage mechanism is the more appropriate paradigm for the gyroverse.

Finally, the mechanism that explains gravity must also explain inertia, since, the equivalence of inertial mass and gravitational mass implies that both have the same root cause. In addition, it also has to explain mass energy, because it too must have the same root cause. From the previous discussion it is clear that graveons are causing gravity and inertial mass. In addition, because mc^2 is due to the expanding universe and not that matter is condensed energy, it too is inertial. Consequently, the graveons cause all three phenomena, suggesting inertial and gravitational mass are the same. Additional understanding can be garnered from another gyroverse characteristic, the gyrohelix mechanism, which needs to be understood.

INERTIAL MECHANISM

THE GYROHELIX

At the centerpiece of the gyroverse model is the gyrohelix mechanism. In Euclidean space, matter is traveling in the motion-direction at the speed of light. From the outside where it really exists, matter is circulating at the speed of light. This circular motion acts like an immense gyroscope. Just like a gyroscope, it resists any torque to its plane of rotation. Yet, unlike a gyroscope, it also advances as it rotates following a helix path. This high circular speed causes a huge inertial amplification factor proportional to the angular circular velocity and inversely proportional to the radius of the gyrohelix. Figure_4.4 is a schematic and description of an ordinary gyroscope that will be tackled before addressing the gyrohelix. The gyrohelix, operating in higher dimension space, shares many of the characteristics of a conventional gyroscope, besides other important attributes due to its multidimensional character.

THE GYROSCOPE OPERATION

One of the most fascinating mechanisms is the gyroscope. In figure_4.4, the gyroscopic flywheel is rotating around the x-axis with an angular velocity ω. A torque τ is applied with the force couple P around the z-axis, which cause a precession of angular velocity Ω around the y-axis. At first, it appears counter intuitive that a torque around the z-axis would cause a rotation around the y-axis. On further examination, the precession causes the flywheel, which was rotating strictly around the x-axis, to have a small component of rotation around the z-axis with a corresponding decrease in rotation around the x-axis. Consequently, the torque around the z-axis is really causing a rotational component about the z-axis. Thinking of it this way, makes it appear a bit more intuitive. Since the applied forces are perpendicular to the precession motion, no energy is imparted on the system. The direction of the angular momentum of the flywheel has been changed, but not its magnitude.

Figure 4.4 - The Gyroscope Mechanism

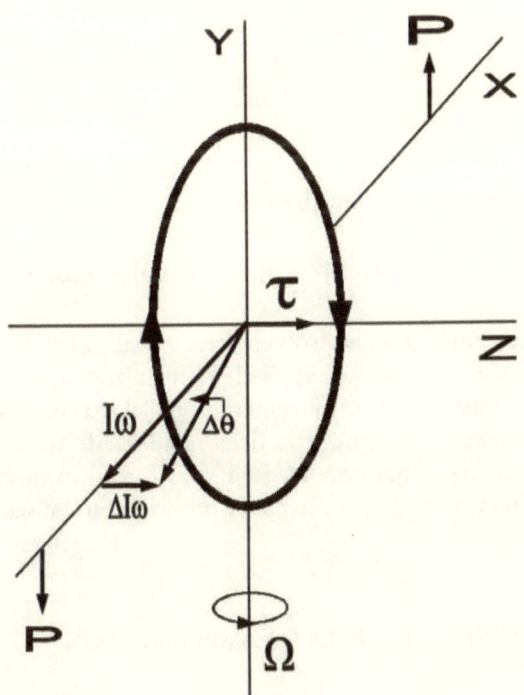

Figure_4.4 –The gyroscopic flywheel is rotating around the x-axis with an angular velocity ω, and I_r its moment of inertia. A torque τ is applied with the force couple P around the z-axis, which cause a precession of angular velocity Ω around the y-axis.

Analytically, for a rotating body, the torque impulse is equal to the change in momentum. This leads to:

$$\tau \Delta t = \Delta I_r \omega$$

For a small angle change $\Delta \theta$

$$\Delta \theta = \Delta I_r \omega / I_r \omega = \tau \Delta t / I_r \omega$$

The precession angular velocity Ω is:

$$\Omega = \Delta \theta / \Delta t$$

Substituting from the previous equation

$$\Omega = \tau / I_r \omega$$

or

$$\tau = I_r \omega \Omega$$

Notice that if the flywheel were moving at lightspeed, and the radius was subatomic, ω would be very large.

$$\omega = v/r = c/.04 \times 10^{-9} = 10^{21} \text{ radians per second}$$

Its effective inertia $I\omega$ is also very large equal to 1×10^{21}. While a gyroscope resists any torque applied to its plane of rotation, it offers little resistance for movement parallel to it. Thus, the gyroscopic effect of the gyrohelix does not impede translation of the plane of rotation, but only resists if it is forced to precess to a new direction. Comparing the type of motion that causes each is interesting.

EFFECTIVE INERTIA AMPLIFICATION

Inertia is such an integral part of everyday life, appearing intrinsic to nature, that it is not noticed. Yet, if looked at critically, it operates very strangely, and is unnatural. For example, Earth has rotated a thousand miles an hour for billions of years unaided, but it would require monumental force to change its rotational speed infinitesimally. That did not happen naturally but was the consequence of a very sophisticated design. The centerpiece of that design is the gyrohelix and its inertial amplification factor.

$$I_e = I_r \omega = I_r c/r = I_r (3)(10^{10})/(4 \times 10^{-11})$$

$$I_e = (I_r) \times 10^{21}$$

Where I_e is the effective precession moment of inertia, I_r is the moment of inertia of matter, ω is angular velocity; c is the speed of light, and r is the radius of the universe atom's rotational motion

The beauty and sophistication of nature's gyroscopic inertial amplification mechanism are overwhelming. Everyone is more familiar than he thinks with this phenomenon, for example, bicycling. It is easy to ride or balance on a bicycle because the inertial momentum of the spinning wheels increases the force necessary to make the bicycle tip-over. If it attempts to tip sideways, the plane of rotation of the bicycle wheel precesses in the direction of the lean. As the bicycle turns into the lean, a centrifugal force pushes out and tends to upright the bicycle and the rider. When going in a straight line, the spinning wheel keeps its plane of rotation in a parallel forward movement, not creating any resistance to this motion. Fortunately, at higher speeds this effect is more pronounced, since the consequence of falling is more detrimental. A simpler demonstration of this effect can be made with a coin. Standing it on its edge is difficult, but when rolling, it easily stays upright.

As discussed, matter in inertial frames at rest or moving at a constant velocity, all travel in its motion-direction. For this motion, the advancing along the gyrohelix moves the plane of rotation parallel to itself, requiring very little force to sustain its motion. An object made to take on a new velocity will invariably change its motion-direction. After the change, its plane of rotation will then move parallel to itself in the new direction, again requiring very little force to sustain its motion. Nonetheless, in going from the first inertial state to the second inertial state, the gyrohelix goes through a period of acceleration. This change in motion-direction causes the gyrohelix's plane of rotation to precess. The force it takes to precess the gyrohelix is amplified by the gyroscopic inertial amplification factor. This factor is proportional to the speed of light divided by the radius of rotation, or about a 10^{19} factor, greatly magnifying the force needed to cause this change. For the non-precessed motion, which occurred before and after the acceleration period, this amplification factor is missing. Inertia has this exact characteristic. Constant velocity causes translation of the plane of rotation and requires little force to sustain it. Acceleration causes precession of the plane of rotation, requiring a large force. This difference alone is very significant in explaining inertia, but an additional factor is explained in the next section, that further increases the difference of the force required for maintaining a constant velocity compared with effecting acceleration.

CREATING INERTIA

The previous section dealt with the fact that changing the plane of rotation takes much more force that translating it. Nevertheless, the energy involved is not just related to the force, but also the action distance for each.

The faster and further matter moves, the larger the force that is developed to impede its motion. For an inertial frame at rest, the gyrohelix's plane of rotation advances 10^{-45} centimeters per revolution or 10^{-27} centimeters in one second. When an object accelerates, it causes the gyrohelix's plane of rotation to precess. A small change of velocity of under ten meters per second, a high automobile acceleration, will precess the plane proportional to the angle $10/c$. Since matter rotates in full space along a diameter of (10^{-9}) centimeters, the precession due to this change of velocity creates a movement of the order $(10^{-9})(10/c)$ or 10^{-18} centimeters per second. Acceleration's precession of the gyrohelix's plane of rotation causes much more movement than a corresponding constant velocity, 9 orders of magnitude more. More motion causes more back force. The important point is that acceleration, which produces a comparable amount of movement in three-dimensional space as constant motion, will cause 9 orders of magnitude more motion in twelve-dimensional space.

Combining this factor with the inertia amplification factor of the gyrohelix's, 21 orders of magnitude, precession produces a tremendous back-force, about 30 orders of magnitude more than constant motion. Another way to view this is that if applying a force in the motion-direction were possible, the force needed to move an object would be reduced by 30 orders of magnitude, because the gyrohelix would not precess.

These two factors together are awesome. The gyroscopic action amplifies inertia, but some is needed before it can be multiplied. The mechanism described here creates inertia. For example, if the velocity were increased to 100 meters per second instead of 10 meter per second, the back force would be 10 times larger. On the other hand, the force generated by just the advance of the gyrohelix would be almost completely invariant to the change. These two methods together create and amplify inertia. Both together, in effect, could create the mass of the universe from the mass of something much smaller than the moon. When the object gives up this energy, the reverse happens. The point is that a very small change in roll (angular velocity), throttles a very large change in energy.

INTRINSIC INERTIA - GRAVEON CONNECTION

The previous two sections showed how a very insignificant amount of inertia is amplified to give objects their substantial inertial quality. Nevertheless, missing was the relationship to the graveon. In this section, the connection is clarified showing how the swarm of fast-moving graveons impinging against the advancing gyrohelix manufactures inertia. It was already shown that precession involves more motion than translation. Graveons permeate uniformly throughout space, traveling across the helix in full space. Objects intercept more of them on the front end than on the back end, tending to slow the objects. The greater the motion, the more slowing, or back force developed. Precession clearly causes more back force than translation. In the last section, it was already shown that acceleration causes 10 orders of magnitude more movement than constant velocity. This translates to 10 orders of magnitude more force.

Another important factor is the pitch, because the graveon force delivered in three-dimensional space is reduced by the leverage-factor, or 36 orders of magnitude. During acceleration, the movement being 10 orders of magnitude larger causes the effective pitch to be 10 orders of magnitude greater. The leverage-factor becomes only 26 orders of magnitude, delivering 10 orders of magnitude more force in three-dimensional space. Both factors together cause 20 orders of magnitude more force.

In the last three sections, four factors were discussed that explain the peculiar characteristic of inertia. However, the numbers cannot simply be summed-up, because of some double accounting. Unfortunately, it cannot be easily unraveled, but the two points to be made are that inertia is intimately tied to the graveon, and matter's intrinsic mass is very small or nonexistent.

With matter not being intrinsically substantial, graveons also do not need as much inertial quality to apply sufficient force to give it these qualities. Consequently, all the mass of the universe may be generated by an orchestrated mechanism of matter, photons, and graveons, each of which, having little or no intrinsic mass. In addition, it is conceivable that graveons and photons are the same particles just moving orthogonal to each other in space.

As an aside, thinking back to the LeSage theory discussion, another reason for its rejection, not previously mentioned, was the concern that all these impinging graveons on matter would generate immense heat on a massive body like Earth, but evidence of it is missing. Because the graveons are generating gravitational forces before matter attains its full inertial quality, heat would not develop.

The full breath of these mechanisms is not easy to quantify, but clearly the universe is a lot less substantial than it appears. An additional factor that reduces further the inherent substance of the universe must await the Cosmology Implications chapter, where the big bang will be discussed.

Inertia and gravitational force both involve the same mechanism of free uniformly distributed graveons permeating space being captured by matter. Matter with twice the gravitational attraction will also have twice the inertia, since both will capture twice as many graveons.

Another way to view this is to recognize that in the LeSage mechanism the combined mass of two objects decreases as these objects come closer to each other, since each blocks more graveons from reaching the other. Because the combined mass decreases their speed toward each other increases to keep the total kinetic energy constant. Hence, their velocities are a function of the change in mass, which implies that the force of attraction between them is also a function of mass. This further makes the LeSage mechanism more compelling.

While the mass of inertia is identical to the mass of gravity, the mass of energy is another mass to consider. A body at rest in three-dimensional space is really rotating at the speed of light, about its plane of rotation, at a helix-pitch of 10^{-45} inches. Thus, matter at rest in three-dimensional space gets its energy kinetically due to its motion, $m_o c^2$. Similarly, all the energy of the universe is derived from its expansion. Since the mass of kinetic energy is identical to the mass of inertia, and matter gets its energy kinetically, it too is the same mass as the other two, making the three identities of mass the same.

THE FOURTH SPACE DIMENSION

During the initial development phases of this theory, after realizing that matter was getting its energy from moving at the speed of light, and not due to matter being energy, the question arose as to which direction it was moving. A first guess was that it was moving in one of the three familiar dimensions, which was on examination not possible.

If all matter were traveling in one of the three positional-dimensions, at the speed of light, the laws of nature would be dependent on direction. Additional motion in the direction in which matter is already moving at the speed of light would combine differently than motion orthogonally to its current motion. For example, in the direction of motion it would add or subtract algebraically. Orthogonal to motion, it would add more like the square root of the sum of the squares. Since the laws of nature are the same in all directions, matter cannot be moving in any of the three positional-dimensions, but instead must be moving in a fourth direction, defined here as the w motion-direction. This has already been discussed.

Figure_4.5 is a geometric two-dimensional representation of the four space dimensions. The vertical axis is the w motion-direction and the horizontal axis represents the three dimensions of position-space. In the twelve-dimensional representation, each vector is in reality a helix. All rest-matter is moving in the motion-direction at the speed of light. If an object moves in position-space at velocity v, relative to the rest-matter, it would cause its motion-direction and positional-dimensions of the object to rotate clockwise, aligning with the rotated c' vector. Vectors associated with the gyroverse are called gyro-vectors, because they represent helixes and combine a little differently than conventional vectors. A velocity in position-space does not make the c vector bigger, but just rotates it. As the motion-direction rotates, its position-space rotates the same amount.

Initially, the object is moving with the rest-matter. In four-dimensional Euclidean space, it is moving in its motion-direction at the speed of light. In the full twelve-dimensional space, it is actually a gyroscopic mechanism, rotating at the speed of light. When an object leaves its rest inertial frame to move in another direction at velocity v, its gyrohelix precesses to a new plane of rotation. The object is still moving at the speed of light, but its motion-direction is slightly different, following the different paradigm for velocity change, as shown in the Figure_4.5. In other words, the motion of an object causes all its axes to rotate slightly such that, it is at rest in its new x', y', and z' positional space, and is moving at the speed of light in its new w'-direction. The vertical c gyro-vector is the original w-direction and the diagonal c' gyro-vector is pointing in the new w'-direction of the inertial frame moving at velocity v. The new x', y', and z'-axis (not shown) are rotated the same amount as the c' gyro-vector.

This discussion is an oversimplification, since inertia holds the vertical velocity constant at the velocity c. This complicates the math, but gives the same result for most real situations, once it is realized that there are two components of the force, one which is horizontally applied and the vertical component due to inertia. Consequently, the velocity of an object including its vertical component is greater than c. the importance of this will become apparent when discussing the speed of light. Nevertheless, for most real situations the vertical component can be omitted. Furthermore, this simplification parallels special relativity formulation more closely.

Figure 4.5 – Gyro-Vectorial Representation of the Fourth Axis (Lorentz Transformation)

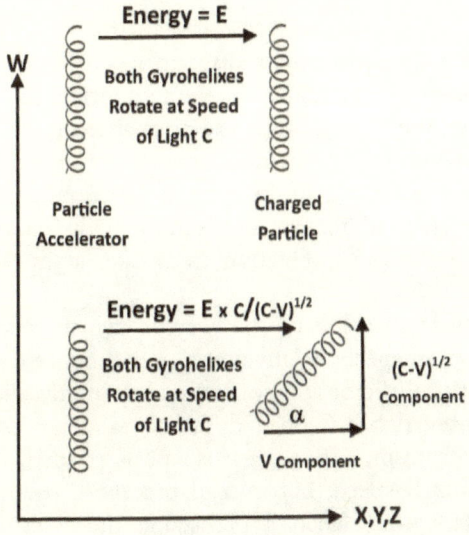

Figure_4.5 - This diagram illustrates the relationship of the w-axis to the three standard x, y, and z positional-dimensions, which are grouped together as one on the horizontal axis. Matter at rest in the standard three dimensions with respect to each other, is moving at the speed of light c as illustrated by the vertical gyrohelixes in the w motion-direction. In the illustration the left gyrohelix is a particle accelerator that is stationary. Particles are introduced into accelerator and starts accelerating. The particles gyrohelix tilts and the fulcrum shortens, requiring progressively more energy to tilt it further, leading to the Lorentz transformation. As the particles go faster the amount of energy needed to tilt it further approached infinity. The charged particles move in the standard three dimensions at velocity v, its gyrohelix precesses to a new direction. For this object, its new w'-axis (not shown) is offset as compared with its original w-direction. The object is now at rest in its new x', y', and z'-axis (also not shown), which is similarly offset from its original direction. Since the motion of most objects are slow as compared with the speed of light, this schematic exaggerates the change in direction of motion for most other practical situations.

RELATIVISTIC TIME AND THE W-AXIS

A review of Figure_4.5 will reveal that this diagram is reminiscent of a special relativity diagram where time, the ct-axis, replaces the w-axis in this model, but with important differences. In special relativity, the units of time are normalized to ct, which is a mathematical device to convert time units to distance units. In this theory, all matter is moving in the motion-direction at the speed of light so that in reality all matters travels in the w-direction a distance equal to ct. It is the fourth direction that makes it appear that time is a dimension.

As previously mentioned, the way in which velocity gyro-vectors combine in special relativity is peculiar because it does not follow conventional vector addition rules. Apparently, these vectors are being combined based on a different paradigm of nature, a gyroscopic mechanism. When a gyroscope is pushed, it does not change it rotational speed, but instead precesses, resulting in this formulation. However, when a gyroscope precesses it does so perpendicular to the force direction, but the gyrohelix moves in the same direction because it involves energy transfer. The fact that velocity vectors in special relativity did not combine as true vectors should have been suggestive that a different paradigm was needed to explain this phenomenon. The gyroscopic mechanism shows that the fourth axis is spatial and not time-like. This diagrammatic drawing for combining velocities is a geometric representation of the gyroverse physical model, not merely a geometric mapping of an equation. Namely, it is a graphical representation of an actual mechanism, and not of just an equation. Notice, if the force was within the particle, or for a rocket ship, the energy needed to accelerate it would not approach infinity.

GYROVERSE ENERGY

Although it takes a great force to precess the gyroscope, it takes no energy, because the force and the precession are orthogonal to each other. On the other hand, for the gyroverse mechanism, the extra degrees of freedom of the gyrohelix, allows energy to be stored and retrieved, because the force can be applied in the same direction as the precession. When a force is applied, the gyrohelix precesses to a new direction, advancing on the gyrohelix, maintaining the motion-direction velocity magnitude. In order for the mechanism to accept the extra energy, the angle alpha is decreased, and the vector approaches the horizontal. Although the torque required to rotate it further remains the same, the force that is also in the horizontal direction needs to increase greatly, because its perpendicular distance to the pivot point shortens. Hence, the product of force times distance (energy) is increased proportionally as the speed increases. Hence, more kinetic energy is stored in the object. When the object's motion is stopped, the process is reversed, and the energy is given up. For gyroscopes, precession does not change energy as in the gyrohelix. Both mechanisms are similar, but the gyrohelix being the more general and the gyroscope being a special case.

The gyrohelix provides the barrier, keeping objects confined to three-dimensional position-space. An object attempting motion to leave this confinement experiences huge forces restricting this motion. To escape this confinement, matter must accelerate. The incremental energy required for acceleration increases progressively, as the velocity increases. It approaches infinity as the speed of the object approaches lightspeed.

These aforementioned objects are further constrained to the Euclidean subspace, since they resist bending in a direction where they have depth. Therefore, once matter in the universe congealed, the structure of the universe was solidified further.

GAMMA FACTOR

GAMMA RELATIONSHIP TO VELOCITY

Gamma (γ) is a measure of the effort it takes to increase the velocity of an object from the rest frame. To derive the gamma relationship of an object as a function of its velocity, refer to the previous Figure_4.5. This figure depicts a situation where an object at rest is moving at the speed of light in a vertically oriented motion-direction. It then takes on velocity v in the horizontal direction. Since v takes place in position-space, its motion is orthogonal to the vertical c gyro-vector. Notice that the velocity gyro-vector v causes the motion-direction to rotate clockwise, around the pivot point, shown as the diagonal c' gyro-vector, leaving the magnitudes of c' and c equal. The velocity v was attained by applying a force to the object to make it accelerate. Actually, it's the torque, force times orthogonal distance (lever arm) to the pivot point that causes the object's velocity. Notice that as the gyro-vector rotates subsequent forces produce smaller torques, by the factor sine α or $(c^2-v^2)^{1/2}/c$. Hence, as the gyro-vector c rotates toward the horizontal, the force required to increase the velocity, or precess the gyrohelix, further approaches infinity. Also note that gravity is just another force, which tilts the gyrohelix.

Expressing this analytically:
$$\text{Force} = \text{mass} \times \text{acceleration} = ma =$$
$$[m_0 a] \times [1/(1 - v^2/c^2)^{1/2}] = \gamma [m_0 a]$$
$$\gamma = 1/(1 - v^2/c^2)^{1/2}$$
$$\&$$
$$m = \gamma m_0$$

This gamma relationship is recognized to be the same as the special relativity gamma function, leading to the Lorentz transformation. Notice that velocity increases the effective directional inertial mass, but not gravitational mass. Inertial mass and gravitational mass are really caused by the same phenomenon. For example, for orbital objects, the inertia mass change is due to the objects speed in the tangential direction, and gravitational mass change is derived by noting its escape velocity in the radial direction. For example, clocks in satellites must be adjusted. Mass is directional. The speed of the satellite increases mass in the horizontal direction, so it slows clocks. The escape velocity of the orbiting satellite is smaller than the escape velocity when the satellite was resting on earth, so its mass in the vertical direction also increases, but that speeds up clocks. The horizontal component is bigger, so the clock slows a little. Hence, the clock periodically needs correction. As the lever arm gets shorter it increases the force and energy required to accelerate an object, not letting it attain the velocity c. This increase in inertial mass also slows clocks. When the velocity decreases, the process is reversed and it gives up this energy. Unlike relativity, gamma is not reciprocal, so inertial mass increases and time clocks consequently slow on the moving frame, no matter from which frame it is viewed. The inertial frame with the lowest gamma is the frame that is stationary with respect to the universe general expansion direction. Alternately, distance shortening is reciprocal by the same gamma factor, but for a different reason discussed later.

As previously discussed, because each axis in four-dimensional space is tightly wound in twelve-dimensional space, each axis that is 14 billion light years long winds to a fraction of an angstrom in length, 36 orders of magnitude smaller. In addition, matter circulating the gyrohelix at the speed of light multiplies the rotational inertia by 21 orders of magnitude, calculated by dividing the speed of light by the helix radius. This is 2 orders of magnitude greater than originally expected because the gyrohelix's radius is 2 orders of magnitude smaller than the original estimate, as will be seen later herein. If not for the gyrohelix, the mass of the largest ship, would be that of a fine grain of sand. In addition, the amount of 12-dimensional motion needed to accelerate as compared to move at a constant speed, adds another 9 orders of magnitude to the inertial mass.

Carrying this analysis a little further:

$$\gamma = (1-v^2/c^2)^{-1/2}$$

$$\gamma = 1 + 1/2(v/c)^2 + 3/8(v/c)^4 + \ldots$$
$$\text{For small v: } \gamma = 1 - 1/2(v/c)^2$$

Substituting into the energy relationship:

$$E = mc^2 = \gamma m_o c^2) = (1 - 1/2(v/c)^2)(m_o c^2)$$
$$E = m_o c^2 + 1/2(m_o v^2)$$

Where E is the total relativistic energy, $m_o c^2$ is the objects rest energy, $1/2(m_o v^2)$ is the object's kinetic energy. Interestingly, the total energy increases, not because the total velocity c increases, but because the inertial mass has increased by the factor γ. This is so because the inertial mass of objects are at least 21 orders of magnitude less during constant velocity than for acceleration, when the gyrohelix changes direction.

Notice that when the gyrohelix c vector rotates, the velocity in the w direction is less than c and does not keep up with the expanding universe. As explained elsewhere, if an object strays from the four-dimensional expanding hyper-surface, the gravity of nearby objects push it back to the hyper-surface.

However, the reason the vertical inertia component can be ignored for this discussion is that for typical maximum velocities of large objects, even asteroids, the w directional inertial stretching will not change v and not appreciably modify gamma. For the speed of light, almost attained by atomic particles, it reaches its final speed long before the gravitational and inertial force kicks in, and does not affect the particles gamma or energy. As mentioned, the force and energy to accelerate an object is limited when the object gets its push from a stationary driver, like a particle accelerator. On the other hand, if the object gets its push from within, as for example, a spaceship, the force and energy limitation does not apply, although the speed limitation "c" persists.

GRAVITATIONAL EFFECT ON LIGHT

Many discrepancies with the equivalence principle could be cited that distinguish gravity from inertia. All of them have been rationalized away by making the principle more restrictive. Keep in mind that except for their differences all things are the same. The elevator example has been described earlier. The fact that gravitational force is graduated, and 'pulls' to a point, referred to as the tidal effect of gravitational fields are other examples. These can't be duplicated in an accelerating elevator. Perhaps the most blatant example is clocks, which slow at a constant rate in a gravitational field, but slow in free space as a function of velocity or the integral of acceleration. The point is that gravity and inertia are distinguishable. Therefore, it would not be strange to discover that photons react differently to these two phenomena. However, because gravity and inertia have the same root cause, it is also not surprising that their reactions are equivalent for many situations. An excellent discussion "On the gravitational deflection on light and particles" by C. S. Unnikrishnan, is referenced in the bibliography, of which the next section calculation is patterned after.

Photons moving in a gravitational field laterally to a massive body, are subject to two distinct, but related deflection mechanisms. If viewed as a particle having a mass of value E/c^2 where $E=h\nu$, it will experience a gravitational force downward orthogonal to it's lateral motion, similar to all other matter, resulting in a deflection angle of $2GM/c^2R$. This is the deflection mechanism that the equivalence principle addresses.

However, because gravity 'pulls' toward a point at the center of mass, there is also a horizontal gravitational component, which squeezes its wavefronts together. The force is greater for photons closer to the massive body, making the wavelength shorter and the front velocity larger. The bottom of the front compresses more than the top, moving faster, and producing bending toward the gravitational source. The bending causes an additional deflection angle numerically equal to $2GM/c^2R$, making the total deflection $4GM/c^2R$, as also predicted by general relativity. An analytical explanation is contained in the next section. For those not interested in the details, it can be skipped.

GRAVITATIONAL EFFECT ON LIGHT CALCULATION

As mentioned, there are two gravitational mechanisms that bend starlight. Imagine a stream of photons nominally moving horizontally at velocity "c" along the x-axis, skimming the sun and heading toward Earth. The Sun's gravitational force can be separated into two components acting on the photon stream, one vertical and the other horizontal. Each causes the stream to bend. However, the correction terms to the nominal paths are so small that they can be calculated without regard to the changes of the nominal path.

Figure 4.6 – Gravitational Deflection of Light

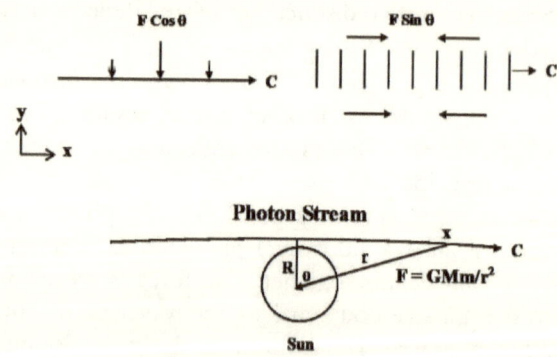

The vertical component deflects the stream downward toward the Sun as it heads toward earth. For simplicity, only the second half from the point of tangency to the Sun, onward to Earth need be considered. The section from the star to the sun is the mirror image, doubling the result. (See above diagram – Upper left))

The gravitational vertical acceleration component on each photon, treated as a particle, is:

(1) $a_y = [F_G/m][\cos\theta] = [GMm/mr^2][\cos\theta]$

where: $r^2 = [x^2+R^2]$
and: R is the distance from the sun's center to the closest point of the photon stream.
and: a_y is vertical acceleration component
and: m is the imputed mass (hf/c^2) of a photon in flight
and: θ is the angle between r and the vertical y direction
and: G is the gravitational constant
and: M is the mass of the sun

The velocity in the vertical direction is the integral of acceleration over time:

(2) $v_y = \int a_y\, dt = \int GM/[x^2+R^2] * [\cos\theta]dt$

where: $[\cos\theta] = R/[x^2+R^2]^{1/2}$
and: $dt = [dx][dt/dx] = dx/c$

Collecting terms:

(3) $v_y = GMR/c \int [x^2+R^2]^{-3/2} * dx$

The angular deflection is:

(4) $a_y = v_y/c = GMR/c^2 \int [x^2+R^2]^{-3/2} * dx$

Integrating:

(5) $a_y = GM/c^2 * x/R[x^2+R^2]^{1/2}$

Evaluating between minus infinity (star) and plus infinity (earth):

(6) $\alpha_{earth} = GM/Rc^2 + GM/Rc^2 = 2GM/Rc^2$

If gravity pulled vertically and not to a point, this would be the total deflection, but that is not the case. A single small particle in twelve-dimensional space is spread out in four-dimensional space. Consequently, the velocity of the photons are not just c, but contains another velocity component, produced by the lateral gravitational force, which causes the photon's velocity to vary with distance from the sun. Consequently:

$$(7) \quad v_x = \int a_x \, dt = \int GM/[x^2+R^2] * [\sin \theta] dt$$

where: $[\sin \theta] = x/[x^2+R^2]^{1/2}$
& v_x is the photons final change in velocity

Since: $dt = [dx][dt/dx] = dx/c$

Substituting and collecting terms:

$$(8) \quad v_x = GM/c \int x [x^2+R^2]^{-3/2} * dx$$

The average fractional change in velocity u_x, noting that the average velocity is approximately half the final velocity, and the nominal velocity is c is:

$$(9) \quad u_x = v_x/2c = GM/2c^2 \int x [x^2+R^2]^{-3/2} * dx$$

Integrating:

$$(10) \quad u_x = [GM/2c^2] \, [x^2+R^2]^{-1/2}$$

Evaluating x between tangent to the Sun and Earth, and doubling to account for the mirror trip from the star to the sun:

$$(11) \quad u_x = [GM/2Rc^2] + [GM/2Rc^2] = [GM/Rc^2]$$

Because photons have wavelike properties, there is a velocity difference from the top to the bottom of the wave front. The velocity gradient is:

$$(12) \quad \Delta u_x/\Delta R = \Delta[GM/Rc^2]/\Delta R$$
$$(13) \quad \Delta u_x/\Delta R = -GM/R^2c^2$$

Notice that this differential is extremely small, but even such a small difference in the compression still causes considerable bending toward the increased front velocity, for example the compressed wave fronts. This phenomenon can be observed in other situations. For example, even small non-uniform shrinkage in a wood plank bends it greatly. For small deflections, the bending can be approximated by a right triangle, where the hypotenuse represents the mid-point photon path, the long side the compressed bottom path, and the small side the deflection. The hypotenuse by deduction is the distance GM/c^2. The long side is the hypotenuse minus $GM/2R^2c^2$. Applying Pythagoras' theorem, the deflection is:

(14) $\beta = - GM/c^2 \, [1^2 - (1 - 1/2R^2)^2]^{1/2} = [1 - 1 + 1/R^2]^{1/2}$

Having dropped the R^4 term & doubling to account for the path from the sun to earth:

(15) $\beta = 2GM/Rc^2$

Finally, since each deflection is small, the total angular deflection can be approximated by the sum of the deflection due to its particle and wave behaviors: (See above diagram – Lower illustration) Since the deflection is small, calculating the bending neglecting the bending in the nominal path is justified.

(16) $\delta = \alpha + \beta = 4GM/Rc^2$

If the bending contributions are compared, it is apparent that while both cause the same total bending, the profile for each is different. Particle bending is greatest while the ray is just grazing the sun. Alternately, the wave contribution is less there, but larger off to each side. This is different from general relativity where both bending components are maximized near the center of mass.

Ironically, Einstein's ubiquitous elevator would exhibit particle bending, with an insignificant wave contribution, resulting in about half as much as general relativity predicts, but the same amount as accelerating in outer space would cause. This is because general relativity's predicted space distortion is also greatest at the point of tangency to the sun, where gravity is strongest and would be expected to maximize in the elevator. Nevertheless, for the entire trip from the star to earth, both deflection contributions complement each other, together deflecting more uniformly than each one separately. Consequently, acceleration and gravity do not cause the same amount of deflection, because for acceleration, the wave component is missing. In the elevator, the bending would be the same for both, but for the wrong reason.

Interestingly, when Einstein, in 1911, originally proposed the equivalence principle, he predicted that light would bend, and proposed that it could be verified during the next eclipse of the sun. The calculation he presented used the wave property, not Newtonian reasoning, apparently assuming that both the wave and particle arguments were not separate and distinct, but just different views of the same phenomena. Consequently, he predicted half the deflection that was ultimately measured. Fortunately, he corrected his prediction when he proposed general relativity in 1916, before the famous eclipse of the sun experiment was finally performed. In the GR theory, the deflection was attributed to space-time warping caused by the massive sun. Space-time warping is not required. This same sort of analysis would work for any small particle.

Also notable is that gravity pulling to a point shorten its wavelength in agreement with the Shapiro delay experiment, in 1964, showing that the wavelength of radar waves bouncing off Venus after skimming the Sun was blue-shifted.

MODIFIED NEWTONIAN DYNAMICS REVISITED

The above Gravitational Effect on Light Calculation (GELC) showed that the correct calculation of light by the Sun can be derived using classical arguments, without resorting to relativity, but using the Gyroverse understanding of its wave like notion of photons. Each point of a photon is spread over many turns of the helix. In addition, it showed that the two components of bending, while each totaling to the same value, unlike present assumptions, have a different bending profile, which is especially beneficial for low acceleration MOND conditions.

Whereas the dark matter hypothesis is the current predominant paradigm, MOND (Modified Newtonian Dynamics), an alternate theory, has significant support. This theory was first proposed in 1983 by Mordehai (Moti) Milgrom, Professor in the Department of Condensed Matter Physics at the Weizmann Institute in Israel. Moti challenged the idea that dark matter was needed far from the galactic center to provide enough gravitational attraction to hold the galaxies in place. He proposed the alternate possibility that the inertial force for accelerating objects might be smaller than Newtonian (or general relativity) theory predicts for slowly accelerating masses. With $F=ma^2/a_0$ for accelerations below a_0, and the Newtonian $F=ma$ only applying to accelerations greater than a_0, he could explain why the stars very far from the galactic center could move so fast, without flying away from the galaxy. For example, if the radius of a star to the galactic center is doubled, the gravitational force is reduced by a factor of four. When orbiting velocity remains the same, as happens for stars far from the galactic center, the acceleration only halves, reducing the outward force by only a factor of two. Under this circumstance the outward force would overpower the force of gravity, tearing the galaxy apart. However, if the force were proportional to the square of acceleration, instead of just simply proportional directly to it, the outward force would also decrease by a factor of four, counterbalancing exactly the gravitational force. This could explain, according to Moti, the galactic formation and containment without resorting to the dark matter assumption, which dates back over 100 years. Moti estimated a_0, the transitional acceleration to be about $1.2*10^{-8}$ cm/sec^2, a very minute rate of acceleration. At that acceleration, it would take roughly 75,000 years for an object starting from rest to build up to the speed of a jet aircraft. At the exact acceleration a_0, the transition point between both formulas, the centrifugal force for both have the same numerical value, so the transition is continuous. However, in reality the transition is probably even more gradual allowing some of the derivatives to also be continuous.

The MOND theory has very compelling characteristics and standing alone would probably have much wider acceptance. The thought that approximately 80% of the matter in the universe cannot be observed, is discomforting; MOND gets around that. Unfortunately, MOND is very inconsistent with general relativity in that it contradicts the equivalence principle, the underpinning of general relativity, which has gravity and inertial acceleration as indistinguishable locally. Less significantly, it is also contradictory to Newtonian Physics (and relativity), where F=ma exclusively. However, the MOND effect would easily be missed, because with earth's rotation, objects at rest on earth already have accelerations far beyond MOND conditions. In the grand scheme of things, the cosmological theories are subordinate to quantum physics or relativity. Even advocates of the MOND theory are somewhat reluctant to pervert general relativity in support of it.

While MOND explains accurately the matter distribution of the visible mass for galaxy rotation, it doesn't predict the lensing of starlight accurately for a star far behind a galaxy. Even using general relativity correct prediction that light bends twice as much as Newtonian theory suggests, MOND still barely corrects enough with it, having a factor of two shortfall with bullet clusters. In the GELC article in the Nov/Dec Infinite Energy magazine, it was argued that lensing bends light twice as much as traditionally expected without resorting to relativity, which would help MOND, without needing a quasi-relativistic modification. However, it would still be off by a factor of two for bullet clusters, except for one thing. The light bending due to GELC is made up from two gravitational components, as gravity effectively pulls to a point, the center of gravity. One component, the horizontal component, acts in the direction of the beam of light, whereas the other component, called vertical component acts perpendicular to the light beam. While both components create the same total amount of bending for Newtonian dynamics, the component in the direction of the light beam produces more bending under MOND than general relativity would. This boost in the bending would work toward correcting MOND's bending shortfall.

When the acceleration is very close to the threshold value, a_0, MOND provides little benefit, but when the acceleration is considerably smaller than the threshold value, MOND provides significant benefit. The vertical force and its accompanying acceleration is maximum closest to the sun where MOND conditions are weakest. On the other hand, since the force and acceleration where the horizontal force component is maximum, is off to the sides further from the sun, it gets more benefit from MOND, because the acceleration small. Consequently, for MOND low acceleration conditions, the horizontal component contributes more bending than the vertical component forces. This works to boost the bending, somewhat compensating for the bending shortfall. Furthermore, the greater distances to light beams traveling between the bullet clusters increases further the disparity between the horizontal and the vertical components and would further reduce any bending shortfall. If that's not enough, there could be differences in the MOND equations for the two acceleration components, which would provide another degree of freedom for enabling MOND to match the actual lensing even still better. There is a synergy between MOND and GELC, each making the other a bit more likely. While hopefully two different MOND acceleration components would not be needed, Einstein/Newton lensing dynamics only works well because of the luxury that undetectable dark matter can selectively be added, virtually, wherever needed.

MODIFIED CENTRIFUGAL ACCELERATION

While MOND seems to come close to explaining the appearance of missing galaxy matter, there is nothing in the gyroverse theory to explain why the centrifugal force should be different for very low acceleration. However, at the speed of rotating galaxy stars the precession tilting of the gyrohelix would create an imbalance of graveon force in the tilt direction, which points the galaxy center of mass, increasing the stars escape acceleration. This could explain why orbital velocities of stars don't decline with distance according to Kepler's third law.

OTHER EFFECTS OF INCREASING GAMMA

Increase in gamma as several effects. The energy of matter increases to $\gamma m_o c^2$, increasing the effective inertial mass m to γm_o. The larger effective mass results in clock time slowing proportionally, since an increase in inertia slows chemical and atomic reactions. Increased rotational velocity of the gyrohelix remembers this increased energy. However, while mass and clock will change proportionally, distances will only appear changed by the same factor, but for a different reason to be discussed shortly. It is the only parameter that is reciprocally changed between inertial frames.

As pointed out previously, when discussing the LeSage mechanism, the gravitational mass increases when the distance between objects increase. Consequently, its change in mass also affects clocks. For example, a satellite's clock slows due its velocity, but speeds-up due to the satellite's greater distance from earth, since it's the changed ratio of the gravitational mass to inertial mass that's proportional to clock speed. Inertial mass is increased in the direction of travel, and gravitational mass is also increased by less graveon shielding in the transverse direction, almost balancing out each other. An object's speed changes its inertial mass, which slows clocks, and shortens distance measurements. The escape velocity is a means for converting gravitational force into units of velocity. It's equivalent to a deficit in velocity, decreasing gravitational mass, and also slowing clocks relative to outer space, but speeding up clocks and lengthening distance measurements in the radial direction, relative to earth's surface. Most importantly, changes to gravitational mass and inertial mass are the same phenomenon, but usually in orthogonal directions. Since gravitational potential is many to many with escape velocity, it isn't gravitational potential perse that is causing changes to clocks and distances. It is the escape velocity.

General relativity equivocates gravity which causes tidal forces, with acceleration, which is parallel and uniform. Reconciling this is what causes very curved space in intensive matter regions. On the other hand, the escape velocity in the gyroverse plays the same role as the equivalence principle in general relativity, except it's exact and doesn't lead to curved space, making it is easier to deal with it.

It is interesting to note that because position space x, y, and z is always orthogonal to the motion direction w, any velocity in position space always increase gamma and consequently mass. It also tilts the motion direction towards the velocity, so that the resulting position space x', y', and z' is orthogonal to the new motion direction w'. Consequently, any new velocity, which can only be in the primed position space, must also increases gamma and consequently mass. It also tilts the motion direction towards the new velocity, and so on. Accordingly, every velocity of an object in open space only increases the objects gamma, energy, and motion direction, but leaving "c" unchanged. However, friction would bring velocity back to zero, and taking everything back to its unprimed state, but at a new location in space.

CHAPTER 5 - RELATIVITY IMPLICATIONS

NON-SIMULTANEITY

EINSTEIN'S TRAIN THOUGHT-EXPERIMENT

Einstein, the thought-experiment master, proposed the train thought-experiment to justify the notion of non-simultaneity. Unlike quantum theory, strange behavior is not associated with relativity. This is probably because distinguished physicists have never pointed out the inconsistencies of relativity, as with quantum theory. Einstein and Schrödinger, who themselves made important contributions to quantum theory, pointed out many inconsistencies in it. To the contrary, some authors try to give the impression that relativity makes perfect common sense. Nothing could be further from the truth! Of the several anomalies discussed in the Problems and Implications chapter, few are as strange as non-simultaneity.

Motion in the observer's w-direction is always at the speed of light, and motion in its x, y, and z directions are always zero. When considering all four directions, objects, especially light, can be moving at a relative velocity higher than c. Although in position-space of the observer, this will always appear to be limited to the relative velocity c. Because of the gyrohelix, velocities do not combine by simple vector addition. For small velocities, the speeds most familiar, simple vector addition is a close approximation to what is really happening. Photons, on the other hand, always travel close to the speed of light, so that the vector addition paradigm leads to large errors. This point is expanded upon with the following set of illustration, which explains the gyroverse's interpretation of special relativity's non-simultaneity paradox.

In figure_5.1, the top train car starts moving to the right with the velocity v at time t = 0. At the center of the train car, two photons are transmitted in opposite directions, as shown by the arrows. The lower two illustrations show the train at time t = T. The middle train illustration is from the perspective of the ground observer, and the lower train illustration is from the perspective of a train observer. From the perspective of both observers, the photons are moving at the same speed of light, but each observer sees the situation differently. What follows are two explanations for the observations; the first is based on special relativity, while the second is based on the gyroverse model.

Figure 5.1 - Simultaneity I - Special Relativity vs. Gyroverse Relativity

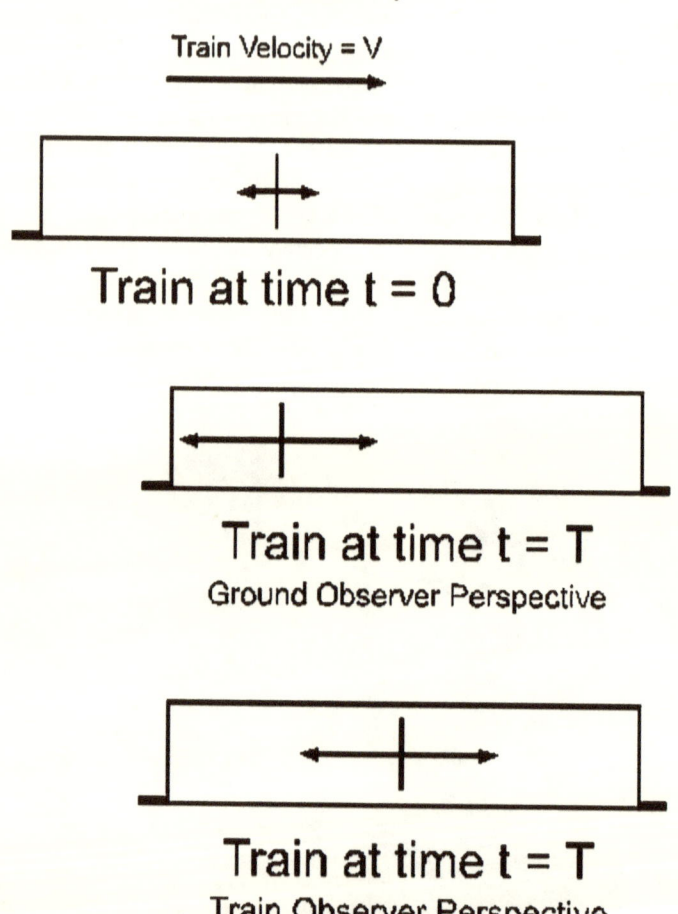

Figure_5.1 - The top train car is moving to the right with the velocity v at time t = 0. At the center of the train car, two photons are transmitted in the opposite direction, as shown by the arrows. The lower two illustrations show the train at time t = T. The middle train illustration is from the perspective of the ground observer, while the lower train illustration is from the perspective of a train observer. From both observers' perspectives, the photons are moving at the same speed of light, but each observer sees a different view of the situation.

Special Relativity: From the viewpoint of the train observer, both photons clearly reach their respective ends simultaneously. Special relativity posits that the speed of light is the same with respect to all inertial frames. Having much less distance to travel from his perspective, the ground observer sees the photon headed to the back of the train car reaching it much sooner than the photon headed to the front of the car. The arrows on the middle diagram illustrate this. Special relativity resolves this apparent paradox by concluding that simultaneous events in the train inertial frame are not simultaneous from the ground inertial frame perspective, by an amount just enough to explain the difference in timing from both perspectives.

Gyroverse Model: From this model, the train observer's perspective is analogous to the special relativity explanation in that the photons strike both ends simultaneously. From the perspective of the ground observer, the situation is different, as shown with the aid of the following two gyro-vector diagrams, representing the photons being fired at the train's front and back walls. One photon is moving forward at the velocity $c+v$, while the other is moving backward at $c-v$, because the photons are moving relative to the train with velocity v. Both photons move at different velocities, but strike both ends of the car simultaneously. The ground observer is in a different position-space than the train observer and does not experience each photon's v component, which is the photon's relative speed to the train, seeing the speed of both photons as equal. Though both photons appear to be moving at the same velocity, the one going further is really traveling faster relative to the train, by just enough to strike their respective ends simultaneously.

Although the relative velocity of the ground appears to be exactly opposite to the photon's direction, they are slightly offset from each other, just enough to hide the v component of velocity.

Figure 5.2a - Simultaneity II - The Gyro-Vector Diagram – Photon and Embankment Moving in Opposite Direction

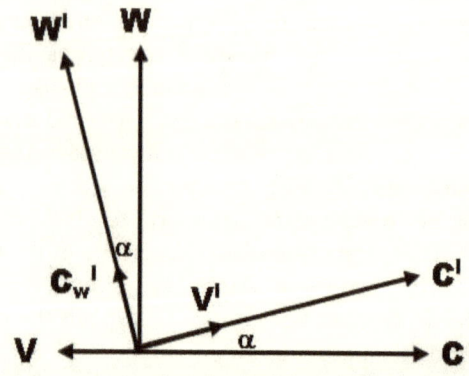

Figure 5.2b - Simultaneity II - The Gyro-Vector Diagram – Photon and Embankment Moving in the Same Direction

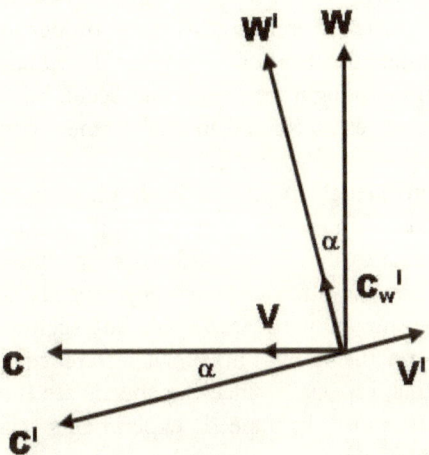

Figure_5.2 - These two diagrams illustrate vectorially the mechanics behind the non-simultaneity illusion of the train example. The diagram is approximately to scale; with the speed of the train in this diagram assumed to be 20% of the speed of light. Non-primed coordinates represent the inertial plane inside the train car and the primed coordinated represent the embankment's inertial frame. The top diagram represents the photons transmitted to the right in the direction of the moving train; the lower diagram represents the photons transmitted toward the rear of the train.

W' is the speed and direction of the embankment, which is rotated slightly due to its v component. The position direction is also rotated, so that the light signal has two components, $c_{w'}$ that is in the direction of W' and is hidden; whereas c' has the same magnitude in both directions. It now appears that getting to the front wall will occur more slowly. Actually, in the total four-dimensional space the photon is going forward at c+v, and backward at c-v, arriving at both walls simultaneously. On the embankment, only the primed coordinates apply. Vector v is zero and c', the projection of c, is only slightly smaller than c. Thus, the speed of light on the embankment is slightly smaller than on the train.

The inertial frame where the photons originate is the only inertial frame where the real speed of light and the measured speed of light are the same. Therefore, this is the proper inertial frame to measure time intervals or distance traveled. In other inertial frames, a Doppler shift in the light frequency will occur, but it is possible to work backward to establish the inertial frame where the photons originated.

Because the actual speeds of both photons, relative to the embankment are different, a Doppler shift results. The rotated position coordinates hide the difference in speed of the two photons, but not the shift in frequency. Speeds of the photons relative to the embankment are equal, but slightly less than the speed of light on the train. This small difference $(c^2-v^2)^{1/2}$ would not be noticeable, especially since the speeds are the same in both directions, and most lightspeed experiments can only make comparative measurements.

SPEEDING BULLET SUBSTITUTION

The previous construction shows how the speed of light can remain constant, viewed from any inertial platform, while simultaneity is preserved. One would normally expect that the relative speed between objects moving in opposite directions to be the sum of both velocities, and the velocity of objects moving in the same direction to be the difference of both velocities. That is the way it is for all conventional motion. To validate the construction, insuring it was not rigged to illustrate the point, it is instructive to test for typical speeds. The following two figures, 5.3a and 5.3b, do just that. Instead of shining light in both directions, a bullet is fired in each direction.

In both diagrams, B is the velocity of the bullets, W_b is the bullet's motion-direction gyro-vector traveling at the speed of light, and D' is the relative speed of the train and the bullet. To make a suitable diagram the speeds of the train and bullet are much larger than normal. Nevertheless, notice that D' gyro-vectors both appear to add and subtract the velocity of the bullet and train correctly. An important point is that these diagrams track the actual physical actions and are not merely vector representations of the equations.

Figure 5.3a - Simultaneity III - Bullet and Embankment Moving in Opposite Direction

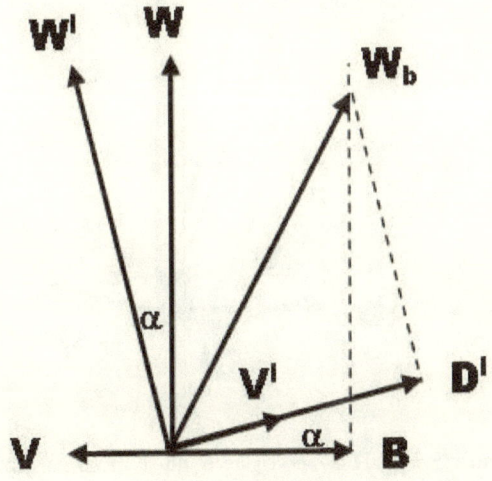

Figure 5.3b - Simultaneity III - Bullet and Embankment Moving in the Same Direction

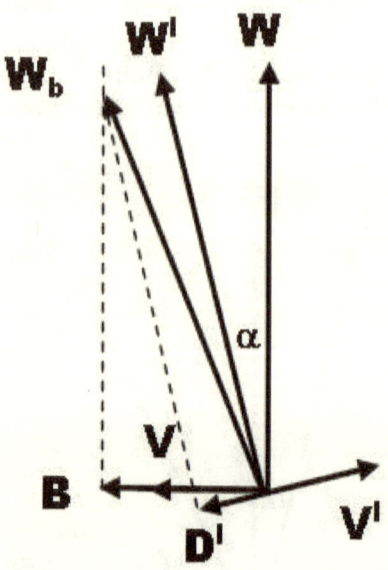

Figure_5.3 - This set of two illustrations is similar to the previous pair, with the difference that bullets replace photons. Since bullets do not travel as fast as light, the bullets cause the relative velocity gyro-vectors v and its value in the motion-direction W_b. Non-primed coordinates represent the inertial plane inside the train car, and the primed coordinated represent the embankment's inertial frame. The top diagram represents the photons transmitted to the right in the direction of the moving train. The lower diagram represents the photons transmitted toward the rear of the train. The B gyro-vector represents the bullet. D' is the projection of the bullet motion-direction gyro-vector on the embankment's position-space.

The B gyro-vector represents the bullet's velocity in position-space. W_b represents the bullet's motion-direction gyro-vector, which does travel at the speed of light. The projection of this gyro-vector, D', on the embankment's position-space, results in a gyro-vector that is approximately the algebraic sum of B and V in

Figure_5.3a, and their difference in Figure_5.3b. For typical train and bullet speeds, this approximation is extremely good. However, at normal train and bullet speeds, the velocity vectors would be too small to be discernable on this scale drawing. Nevertheless, at these intermediate speeds, the sum and difference approximations are still very good. As the bullet's speed approaches the speed of light, D' approaches W_b, the speed of light, duplicating figures 5.2a and 5.2b. Consequently, this physical construction has the requisites to explain all motion from typical to relativistic.

SPEED OF LIGHT – BINARY STARS

Non-simultaneity was proposed by special relativity to explain the fact that the speed of light appears the same independent of the relative speed of the emitter and observer. This seemed to be the only way to explain this strange behavior. The gyroverse theory presents a different explanation. This apparent paradox emerges because the geometry of the universe is not what it seems.

The previous thought-experiment to explain this phenomenon probably could not be carried out with today's technology. Actually, the important aspect of the experiment is to show that the constancy of the speed of light is not related at all to any strange nature of space-time, but simply to the construction of space. Fortunately, certain binary stars, originally studied by Willem de Sitter can demonstrate this odd behavior.

Binary stars are two stars that are gravitationally bound to each other, orbiting about their center of mass. Roughly, half the stars in the sky are in these binary systems. Some of these binary systems consist of a normal star orbiting with a neutron star or a small black hole, close enough together to be moving at a significant fraction of the speed of light, and completing an orbit in less than a day. The stronger their gravitational attraction, the greater their orbital speed need be to keep them separated. The following figure illustrates this arrangement.

Figure 5.4 - Binary Star System

Figure_5.4 - This is a binary star system, in the plane of Earth, orbiting around their center of mass, often completing the cycle in less than a day. The black star is a neutron star or small black hole emitting no light, and the white star is a normal star. Because of the general universe expansion, light from the binary star system is red shifted. Superimposed on this overall ambient red shift is a blue shift when the star is moving toward Earth, and an increased red shift when it is moving away. Because it is moving forward for as much time as it is moving away, the absence of a difference in time between the two periods is an indication that the speed of light is constant for both periods.

The significance of this arrangement is that only the normal star emits light. Actually, the x-ray spectrum is utilized to circumvent the extinction effect. When the star is moving toward Earth, its light is blue shifted; and when it is moving away, its light is red shifted. If light moves slower when red shifted, that period would be longer than the period where it is blue shifted. Doppler shift periods being perfectly symmetric indicate that the speed of starlight when the star is moving toward or away from Earth is the same. This shows that the speed of light is constant, independent of the relative speed of the source and the receiver. However, in the four-dimensional space, the speed toward or away of light is the sum or difference of the speed of light and the speed of the star. Consequently, the doppler shift in four space dimensions carries over to three space dimensions, but the speed difference of light and star does not.

The next set of gyro-vector illustrations, figures 5.5a and 5.5b, explains vectorially why this is so.

**Figure 5.5a - Binary Star System Gyro-Vector Diagram
(Star Moving Away From Earth)**

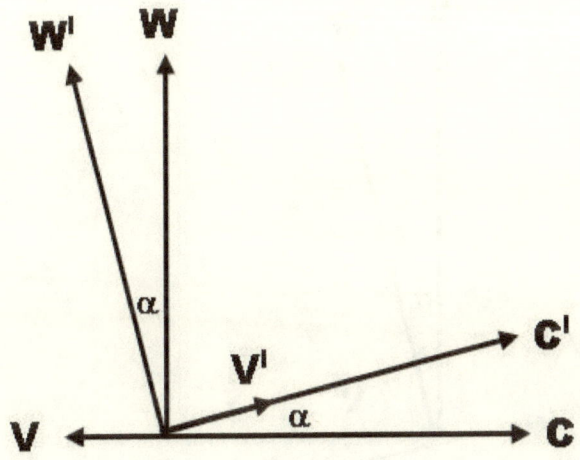

Figure 5.5b - Binary Star System Gyro-Vector Diagram
(Star Moving Toward Earth)

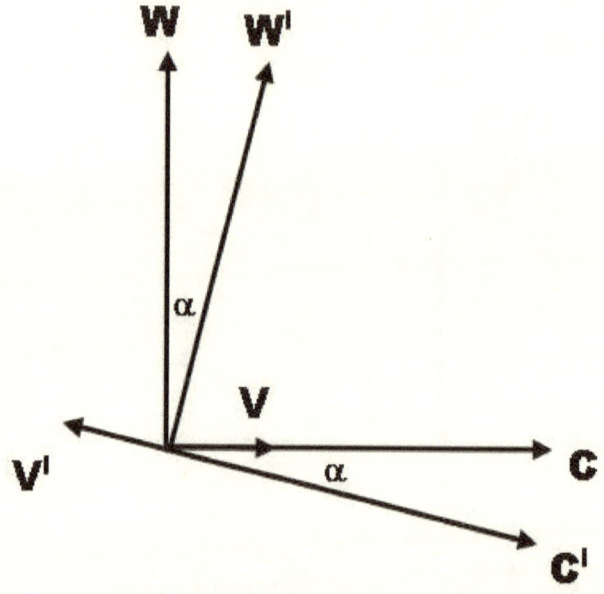

Figure_5.5 - The unprimed coordinates represent the rotating star's inertial frame, and the primed coordinates represent Earth's inertial frame. The top illustration represents the inertial frame in which the star is moving away from Earth. The lower frame represents the star moving toward Earth. Vector v is the relative speed of the star and Earth, and W shows it in the motion-direction. Notice, in both cases, the speed of light c' is identical, at least to a first approximation. Again, this apparent constant value of light is just due to the fact that position-space for a different inertial frame is different by just the correct quantity to hide v.

These illustrations are exaggerated, and not to scale. The unprimed coordinates represent the rotating star's inertial frame; the primed coordinates represent Earth's inertial frame. Vector v is the velocity of Earth relative to the star. Vector v' is the velocity of the star relative to Earth. It is the projection of the star's motion vector W on Earth's position-space. It is the same magnitude as v, but not exactly opposite to it, since both position-spaces are different. Gyro-vector c is the speed of light from the star to Earth. Vector c' is not a projection of it on Earth's position-space, but is equivalent to c. Unlike the train thought-experiment where the light rays are confined entirely to the train's inertial frame, this example bridges two inertial frames. The light ray is emitted from the star, but is captured by Earth, two different inertial frames, celestial distances apart. Therefore, in both inertial frames, the light ray travels at light's characteristic speed.

The top illustration, figure_5.5, represents the inertial frame in which the star is moving away from Earth, while the lower frame represents the star moving toward Earth. Notice that the speed of light on Earth is identical in both cases. Again, the apparent constant lightspeed occurs because position-spaces for the different inertial frames are rotated just enough to hide v.

Alternately, because the star is so far from earth the light from it essentially travels along the same straight line to earth independent of its motion. Consequently, the gyrohelix vector keeps it speed toward earth constant independent of local motion.

GYROVERSE CHARACTERISTICS

YAW, PITCH, AND ROLL

The universe's gyroscopic mechanism, gyrohelix, is the major controlling factor in the operation of the universe. All matter is moving at the speed of light along the helix. This effect is controlled by three variables, names borrowed from the nautical:

the yaw, the direction perpendicular to the plane of rotation of the gyrohelix, the pitch, the longitudinal advance of the plane of rotation of the gyrohelix for each revolution, and the roll, the angular velocity of the gyrohelix. During the initial phases of the universe, while the expansion was still accelerating, the r/p ratio was constant, but their r-p product increased from zero to its present value. Once the universe completed its initial accelerated expansion, the product has remained constant. These three parameters can be viewed from the unwound perspective, the way we experience the universe, or from its twelve-dimensional perspective. Keeping both perspectives in mind is useful since different physical attributes are better understood from one or the other. All neighboring matter in our galaxy is moving at essentially the same yaw, pitch, and roll. In the Cosmology Implications chapter, a diagram and discussion will be presented showing how our Milky Way fits within the rest of the galaxies in the universe. It suffices to say that all galaxies are moving at essentially the same pitch, and roll. In contrast, the yaw is uniformly distributed over all four dimensions on a 3-sphere.

Usually, when one object accelerates relative to its previous rest state, it causes a gyrohelix precession, changing its yaw. Distance and mass increase, and time dilation are caused just by the change in yaw. The tricky part is that two objects in relative motion view each other's shortening of distance in their respective motion-direction due to their different yaw. Consequently, the mechanisms for distance change only depends on the viewer's perspective. The shortening due to yaw is symmetrical from both objects' point of view, but only occurs in the direction of relative velocity, because they each can only view the component of the other object's motion in their own position-space.

There is no limitation on matter traveling faster than light. The limitation is only that the force required would approach an astronomically large number as velocity is increased toward the speed of light that limits this velocity. If it were possible to increase the speed without the accompanying precession, speeds above the speed of light could be attained with a small force. Later in the book, the universe expansion mechanism will be discussed where this change occurs, without a corresponding change in yaw. An interesting consequence of this is that it takes much less energy to do it—at least 28 orders of magnitude less. When the energy is reclaimed, as in nuclear fusion in stars, a change in yaw results, which produces the full amount of energy. Therefore, though less than one percent of the matter is destroyed in the process, 27 orders of magnitude more energy is created as it took to produce all the mass energy in the first place. In effect, celestial objects have more energy than originally imparted to it.

In addition, a change in yaw is possible without a corresponding change in mass or energy. If the direction, but not the magnitude of velocity changes, only the yaw changes, and energy is not added or subtracted. One example is with a perfectly elastic collision of a light object and a very heavy object. The light object bounces off in a new direction at the same speed. Another example is an object on the surface of Earth. Its yaw continuously changes, repeating each time Earth completes a revolution. Consider two timepieces on the Equator, but on opposite sides of Earth. They would have a relative motion of 2 thousand miles per hour. Because of the symmetry, neither timepiece could be running slower, even if the timepieces were monitored from only one site. The fixed distance, from one site to the other, would keep the delay of transmission and the time lag a constant. A time loss could only occur if one timepiece consistently ran slower than the other did.

On the other hand, consider two timepieces, one at the Equator and one at the North Pole. The Equator timepiece having more kinetic energy and greater gamma, will run slower wherever time is monitored. Unfortunately, this difference is probably too small to be measured.

DISTANCE

As just previously mentioned, if an object is moving relative to a stationary platform, its kinetic energy is increased. This will increase its gamma, and correspondingly increase its effective mass and energy, and slow its clocks. From the rest frame, the moving object would appear foreshortened, in its direction of motion for a different reason. A moving object is in a slightly offset position-space compared with the rest object. The component of motion in the rest object's motion-direction is hidden from the rest object's view. It is somewhat similar to an object moving obliquely away from a viewer appearing shortened. Here, the "away direction" is the rest object's unseen fourth dimension. Similarly, the rest object will also appear shortened to the moving object for the same reason. From each object's perspective, the other appears foreshortened in the direction of relative motion. Conversely, all distances orthogonal to their relative motion remain unchanged.

TIME, A BYPRODUCT OF INERTIA

While none of the dimensions of this model is time per se, clock time is also slowing proportionally to the inverse of gamma. It is not time that is changing, but just all the measures of time that slow. The increase of inertia, as gamma increases, effectively slows all atomic reactions, including atomic timepieces. The larger inertia would also slow all mechanical timepieces and make all motion more sluggish. Living processes would slow, but whether they would slow in proportion to physical processes is not clear. Nevertheless, the slowing of time and biological processes would tend to keep things appearing and feeling natural, counterbalancing the sluggishness introduced by the higher inertia.

Creating the conditions necessary to feel the effect of increased inertia is not presently possible. Despite this, an analogy exists between small quick animal species and large slower moving animals. Small animals have more strength relative to their body weight than larger animals. The larger animals mimic the actions of smaller animals under high inertial conditions. Overall, the larger slower moving species live much longer than the small fast-moving species. In effect, their internal clocks are moving slower. A surrogate marker for this comparison is heart rate. The number of heartbeats over a lifetime tends to be more representative of the life span for different species than the number of years they live. An elephant lives much longer than a mouse when measured in time, but about as long when measured in total heartbeats. Slower heart rates are indicative of slower biological processes and consequently longevities. Another interesting example of this phenomenon occurs in humans as they age. Children tend to have high energy level and a high heart rate. Their internal clock runs fast, but they see the adult world and time as moving slower. Typically, children believe they will never grow up. Seniors, quite the opposite, slow down, are more forgetful, and experience the world and time, as moving faster. Andy Rooney's quip captured it best. "Life is like a roll of toilet paper. The closer it gets to the end, the faster it goes."

MASS AND ENERGY

The effective mass and energy of matter in motion is increased in proportion to the increase in gamma. The inertial frame at rest with the universe's expansion would have a true gamma of one. All other inertial frames would have higher gammas. Nevertheless, the inertial frame at rest on earth actually has a larger varying gamma, but is renormalized to have a gamma of one, since all other inertial frames are compared against it. However, any increase tends to escape detection because all measuring standards would increase proportionally. Nevertheless, when viewed from a different inertial frame, the mass and energy differences can be observed.

SPEED OF LIGHT

There are several aspects of lightspeed that need understanding. There are two velocities associated with photons, as well as all other matter. The first comes from the general expansion of the universe in the local motion-direction. It is hidden from observation, proceeding at the speed of light. It will be discussed further in the Cosmology Implications chapter. The second motion occurs in position space, where we observe it. A question arises as to why light and matter, which current theory proclaims to be different should have the same speed limitation.

The reason is simply that photons are just tiny chips of matter kicked with so much energy that its velocity approaches the speed of light within position-space. However, in real sense, photons are moving faster than the speed of light in an oblique direction to both motion and position-space. It will be described more fully later in the treatise.

Another issue is the mechanism by which objects have a maximum speed. As described earlier, when the gyrohelix rotates toward the horizontal, it takes increasingly larger force and energy for objects to approach the speed of light.

Finally, when observers move relative to objects, the extra dimension, not seen, works to hide the extra relative speed beyond lightspeed's threshold. The universe expansion with its gyrohelix mechanism is the root cause for the speed of light velocity limit.

LORENTZ TRANSFORMATION

The Lorentz transformation was originally developed to reconcile the experimental speed of light anomalies by showing that distances shrink in the direction of motion just enough to explain the constancy of the velocity of light in all directions. The theory also predicted a mass increase and time dilation needed to explain some experimental results. These phenomena were explained by the assumed effect that the aether pressure would have on moving objects. The rationale was unconvincing. In effect, the equations were predictive, but the explanations of the physics principles appeared contrived. Einstein, regarding special relativity, took the same set of experimental results, the same Lorentz transformation, but by rejecting the existence of aether, could re-derive the transformation and developed a more convincing relativity theory by attributing the constancy of the speed of light to nature, not having to explain the mechanism. Again, the Lorentz transformation is being used--this time with the gyroverse model. In the gyroverse model, the Lorentz transformation is just an approximation. It is being presented in this form to make comparison with its predecessors easier. While each theory essentially uses the same transformation, each embodies unique variations. Special relativity presumes that all inertial frames are equivalent, and that time slows, and distances shorten with speed in any other inertial frame. Lorentz relativity, on the other hand, assumes the local inertial frame is special, and time clocks slow and distances shorten on a moving frames relative to it. The gyroverse relativity is somewhere in between, in that time clocks run slower on moving frames, but distance shortening is reciprocal between any two inertial frames, like special relativity. However, the CMB isotropic inertial frame, the frame at rest with the universe expansion direction, is the most special, having a minimum gamma. Nevertheless, they all give very similar numerical results in most practical cases, exemplifying that very different physical constructs could lead to the same formalism. This demonstrates that mathematics alone is not nearly sufficient to determine how nature behaves.

Although special relativity evolved to become widely accepted, Lorentz never did quite accept it. It appears that he needed to have a physical reason distances shortened and time dilated. Lorentz could not accept it as an unexplainable quirk of nature. He never completely abandoned the aether hypothesis because he could not contemplate any other physical explanation, though apparently there was no aether. Thus, a parallel between his thoughts on special relativity and Einstein's disbelief of the "action at a distance" aspect of quantum mechanics can be drawn. Both were able to see the illogical nature of theories they were not attached to, yet had a blind spot for ideas very dear to them. Another physicist, Erwin Schrödinger, who developed the wave function, could be added to this group. He believed that the wave function expressed the behavior of physical matter. One of the more prominent physicists then, Max Born, Nobel Prize winner and originator of the phrase quantum mechanics, reinterpreted the equations as a measure of probability. This latter interpretation, never accepted by Erwin Schrödinger, continues to be the dominant explanation.

The following equations describe the Lorentz transformation for this model as viewed from the rest inertial frame, the space-platform. It is only presented as a reasonable three-dimensional approximation to the actual transformation, which previously has been discussed in more detail and is based on the twelve-dimensional gyroverse model. Two identical space vehicles orbiting earth in opposite directions would have the same gamma, mass, and clock speed. However, each would view distance shortening on the other in the direction of separation. The primed variables are the coordinates on a spacecraft that is moving at velocity v relative to the space-platform whose parameters are identified as unprimed. Velocities v, and v' are the velocities of the spacecraft and reconnaissance vehicle relative to the space-platform, and spacecraft respectively. Velocity w is the velocity of the reconnaissance vehicle relative to the space-platform.

A way to view the situation is to imagine that the unprimed coordinates are on a space-platform near Earth, and the primed coordinates are on a spacecraft traveling at a large fraction of the speed of light in the plus x-direction at velocity v. From the spacecraft, a smaller reconnaissance vehicle is heading further in the plus x'-direction at velocity v' relative to the spacecraft. A monochromatic electromagnetic beacon on the spacecraft is beamed in the minus x' (or minus x) direction toward Earth's space-platform. For this situation, the set of transformation equations that convert events relative to the spacecraft (primed coordinates) to conditions relative to the space-platform (unprimed coordinates) is:

LORENTZ TRANSFORMATION CHART - GYRO

Coordinates	$x = (x' + vt')/\gamma$ $\Delta x = \Delta x'/\gamma$ $y = y'$ $z = z'$
Time Dilation	$t = \gamma t'$ $\Delta t = \gamma \Delta t'$
Velocity Addition (Transformation)	$w = [v] + [v'/\gamma]$ $w = [v] - [v'/\gamma]$
Mass, Momentum, & Energy	$m = \gamma m'_o$ $p = mv = \gamma(m'_o v) = \gamma p_o'$ $E = mc^2 = \gamma(m'_o c^2) = \gamma E'_o$
Doppler Effect	$f = \gamma f' (1-v/c)$ $T = T'/\gamma(1-v/c)$
Gamma	$\gamma = (1-v^2/c^2)^{-1/2}$

COORDINATES

The first four equations express the coordinate changes between the two inertial frames. Coordinates y' and z' are unaltered because no movement in those directions occurs. The x' coordinate, the location of the spacecraft, is moving away from the space-platform. Its distance from the space-platform anytime is vt'. In addition, because it is moving at the velocity v, all distance measurements in that direction are shortened by the factor $1/\gamma$, which is always less than one. Another way to view shortening of distances in the direction of travel is to realize that the lengths of objects are zero in the w-direction. When an object moves away from a stationery observer, its w-direction is changed. The observed motion becomes a combination of the original position-space where the object is its actual size and the w-direction where the length is zero. This shortens the size of the object in the direction of motion. Observers on the moving object recognize the full size because their position-space is orthogonal to their w-direction. On the other hand, the moving observer will measure a shortening of the stationery observer in the direction of their relative motion for the same reason. Thus, this phenomenon is reciprocal, and due to changing yaw, unlike, for example, the slowing of time, which is directional, and due to a changing gamma. It is this reciprocity for distance shortening, which gives the misleading impression that any inertial frame can be considered the rest frame. In particular, the spacecraft would view the distance from where he came and to where he is going as foreshortened. On the other hand, observers on the space-platform would see the distances as normal, but the length of the spacecraft shortened.

TIME DILATION

The next set of equations is the time (clock) dilation equations relating time of an event on the spacecraft to the corresponding time of that event on the space-platform. Appropriate adjustments are made to account for the speed of light, the usual mode of event notification. These equations convey when events occur, not the time of awareness. Comparing the time dilation equations in this model with special relativity, shown later in the Relativity Overview chapter, notice that the distance term is missing. This term leads to the non-simultaneity consequences of special relativity. Time is much less mysterious in this model. It is simply the slowing of natural phenomena due to increased inertia caused by an object's increase to its mass. When the object returns to the space-platform, its timepiece speeds up again running in sync with the space-platform. Nevertheless, any lost time while the timepiece was running slower remains lost. A person's internal processes would also have run slower, slowing his physical aging.

Time Dilation by Velocity is proportional to
$\quad$ Gamma = $1/(1-v^2/c^2)^{1/2}$
$\quad$ where v is the surface velocity.

Time Dilation by gravity is proportional to
$\quad$ $1/$Gamma = $(1-v^2/c^2)^{1/2}$
$\quad$ where v is the escape velocity, and gamma's are different

The Potential Energy on a celestial body is GMm/R
$\quad$ Where G is the gravitational constant, M is mass of Celestial body, R is its radius, m is mass of projectile

The Escape energy from it is $\quad 1/2\ mv^2$

Setting potential energy to escape energy, and

Solving for escape velocity $\quad v = (2GM/R)^{1/2}$

VELOCITY ADDITION

The following set of two equations is the velocity addition equations (velocity transformation formula). It is different from the special relativity formula. Velocity v is for the spacecraft relative to the space-platform, and velocity v' is the reconnaissance vehicle relative to the spacecraft. The velocity of the reconnaissance vehicle relative to the space-platform is computed with the formula for w.

Note that γ is the gamma of the spacecraft. The velocity of the reconnaissance vehicle relative to space-platform is always less than the v+v', the corresponding Galilean expression. The next figure illustrates velocity addition.

Figure 5.6 - Velocity Transformation

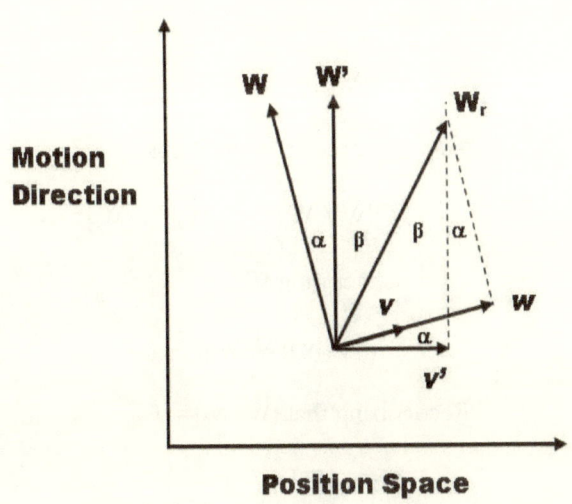

Figure_5.6 - This diagram shows the velocity relationships of the space-platform, spacecraft, and reconnaissance vehicle. The vertical axis represents the motion-direction of the spacecraft. The horizontal axis represents the three standard x', y', and z' positional-dimensions grouped together as one. It is moving relative to the space-platform with velocity v, shown as the gyro-vector orthogonal to the W gyro-vector. The reconnaissance vehicle is shown traveling relative to the spacecraft with velocity v' causing its yaw to precess its motion-direction clockwise to the W_r direction. The projection of the W_r gyro-vector on the space-platform's position-space is the w velocity gyro-vector, the sum of the two speeds. By construction, it is apparent that for typical low speeds they almost add Galilean.

VELOCITY ADDITION INTERPRETATION

Notice that v' only increases the velocity w by v'/γ. If the velocity direction of the reconnaissance vehicle were reversed, then the velocity increment would become $-v'/\gamma$. This velocity reduction is not small enough to explain the Fizeau experiment. To do so, the reduction in velocity would need to be v'/γ^2 and $-v'/\gamma^2$ respectively instead. This velocity increase follows from the equation for w, which is:

$$w = W_r \sin(\alpha + \beta)$$

$$w = W_r (\sin \alpha \, \cos \beta + \sin \beta \, \cos \alpha)$$

$$w = W_r ([v/W'][W'/W_r] + [v'/W_r][W/W'])$$

Factoring W_r

$$w = ([v] + [v'][W/W'])$$

Recognizing that $[W/W'] = \gamma$

$$w = [v] + [v'/\gamma]$$

Nevertheless, it is doubtful whether this or the relativity paradigms are correct for this situation anyway. Light traveling through water is never moving at its average velocity, since it is really made up of fits and starts. When the photons are captured by atoms they cease moving, but when re-emitted they go at the speed of light, for vacuum. Using their average velocity, which special relativity does, is very problematic.

Other effects to consider could mitigate this discrepancy with the Fizeau results. One is the retarded velocity due to the collision, absorption, and reemission of the photons striking atoms in the medium. When the water moves in the same direction as the light ray, the photons must travel further before striking an atom, which tends to speed the photons through it. When the light ray is beamed opposite the water flow, the photons slow down. This explains some of the difference.

Another effect to consider is the Doppler shift (change in wavelength) experienced by each beam of light as it enters moving water. This also reduces the time difference through both paths. Light moving with the flow of water will be red shifted, and light moving against the flow will be blue shifted. Blue shifted light has a higher index of refraction and moves slower through the water.

While special relativity comes closer to matching the experimental evidence without these effects, it is also suspect, because it does not come close enough to insure it really explains the Fizeau results alone, without the need of some additional correction factor.

Finally, observe that when the velocities are small relative to the speed of light, the gammas approach one and the Galilean transformation results.

LIGHTSPEED VELOCITY LIMITATION

It is apparent that a spacecraft's speed can exceed the speed of light, although it might be hidden. However, the particle speed in a particle accelerator is limited by the speed of light. While at first blush the two might appear similar, they are different.

In a particle accelerator all the force for accelerating the particle occurs from Earth's inertial frame, so that all the force is in the same position-space. Under this condition, the force required would approach infinity as the velocity approaches lightspeed.

On the other hand, a spacecraft only applies its force on Earth's inertial frame initially. From that point on, additional force is applied on the inertial platform that the spacecraft attains at each new speed. While subsequent forces needed to accelerate the spacecraft increases, it does not approach infinity at the speed of light. The reconnaissance vehicle in the previous example is an oversimplification of this, where two inertial frames are assumed, instead of it continuously changing. Even then, the speed of light can be exceeded, or at least not require infinite energy. However, stopping when the destination is reached will take as much energy.

RELATIVE VELOCITY OF INERTIAL FRAMES

Referring to the previous Figure_5.6 again, consider the two inertial frames, the space-platform, and the spacecraft. Shown are two gyro-vectors labeled v. Pointing left is the velocity of the space-platform from the spacecraft's perspective. Pointing right, inclined up, is the same velocity, except from the space-platform's perspective. Both have the same magnitude as expected, but they are not precisely opposite each other because position-space of the space-platform and spacecraft are slightly offset from each other. Oddly enough, that is not the full story, since it is not obvious why the relative velocity from both perspectives should be the same. The full reason will be explained later in the chapter with a space travel metaphor

MASS, MOMENTUM, AND ENERGY

The next set of three equations is the mass, momentum, and energy equations respectively. The term E'_o, p_o and m'_o, refers to the mass of an object in the spacecraft. E, p, and m are the energy, momentum, and mass of the object as viewed from the unprimed coordinates or space-platform. Energy E is the total energy of the moving matter, including its mass energy. The energy and momentum equations are shown in several equivalent forms. The energy equation is not recognizable from its Galilean counterpart, since there does not seem to be a "$1/2\ mv^2$" term. Expectantly, as the velocity approaches small ordinary values, the kinetic energy does approach $1/2\ mv^2$. This is discussed more fully in a subsequent section. These parameter changes are directionally transformed. For example, a proton from a particle accelerator would exhibit the energy of an elevated mass; nevertheless, if a human could ride on the particle, atomic reactions on Earth would not appear more energetic to him (but distances on Earth, in the direction of travel, would appear shorter).

KINETIC ENERGY FROM GAMMA FACTOR

The gamma factor is the major factor for expressing the relationship between the rest and the moving inertial frames. Its value is always greater than or equal to one. Unlike special relativity, the gamma parameter is not the same for all inertial frames. The laws of physics are covariant, for all inertial frames, but the value of gamma is larger on the moving frame. Gamma changes all physical quantities proportionally so that it cannot easily be found out by comparing physical parameters in any given inertial frame from within that same inertial frame.

As gamma approaches one, by dropping all of the smallest terms, the Galilean equations will result. The only equation where this procedure is not self-explanatory is in the energy relationship. To get the Galilean energy relation, the gamma term has to be expanded into a series using the binomial theorem as follows.

KINETIC ENERGY DERIVATION

$$\gamma = (1-v^2/c^2)^{-1/2}$$

$$\gamma = 1 - 1/2(v/c)^2 + 3/8(v/c)^4 + \ldots$$

For small v: $\gamma = 1 - 1/2(v/c)^2$

Substituting into the energy relationship:

$$E = mc^2 = \gamma (m_o c^2) = (1 - 1/2(v/c)^2)(m_o c^2)$$

$$E = m_o c^2 + 1/2(m_o v^2)$$

In Newtonian physics, nothing like mass-energy equivalence evolved.
Thus $m_o c^2$, matter's rest energy can be subtracted from both sides resulting in:

$$K_e = E - m_o c^2 = 1/2(m_o v^2)$$

$$K_e = 1/2(m_o v^2)$$

Where K_e is classical kinetic energy, E is the total relativistic energy, v is the object's velocity, m_o is the rest mass, and c is velocity of light.

TOTAL ENERGY IS MC^2, NOT $1/2(MC^2)$

One interesting sidelight is why total energy does not have the 1/2 factor associated with the classical kinetic energy, since the total energy is also kinetic in nature. For example, kinetic energy has the form:

$$K_e = 1/2(m_o v^2)$$

The 1/2 factor comes about because the velocity v increases for increasing kinetic energy. The kinetic energy form is obtained by integrating the following:

$$K_e = \int F dx = \int d(mv)/dt * (vdt) = \int m_o v(dv/dt)dt = \int m_o v dv = 1/2(m_o v^2)$$

where m is a constant equal to m_o.

For the total relativistic energy, the situation is different because the total velocity of motion in twelve-dimensional space, which is the speed of light, does not change. It is the gamma factor that is changing.

$$E = \int F dx = \int d(mc)/dt * (cdt) = \int m_o c^2 (d\gamma/dt)dt = \int m_o c^2 d\gamma$$

$$= \gamma m_o c^2$$

where c and m_o are constants, and $m = \gamma m_o$

As seen previously, when the gamma factor is expanded, the classical kinetic energy equation results for small v. This is a cross check on the appropriateness of the derivations. In simple terms, the total energy of all objects is mc^2, only the component of kinetic energy in the observer's position space is $\frac{1}{2}(mv^2)$.

DOPPLER EFFECT

The Doppler expression relates the frequencies of the beamed electromagnetic wave from the spacecraft to the space-platform. Although the speed of light from any source measures the same in any inertial platform, the frequency of the light exhibits a Doppler shift when the source is from a different inertial frame. Since the station and craft are receding from each other, the frequency is decreasing and the period of the wave, which is the reciprocal of the frequency, is increasing. The Doppler shift only depends on the difference of velocity between the source and the receiver of the light. It does not make a difference which object is moving. This result would be the same for special relativity. Interestingly, the Galilean transformation is different when the source is doing the moving.

GYROVERSE POTPOURRI

STREAMLINE EFFECT

Euclidean space of the gyroverse has four dimensions, the three dimensions of position-space and the one dimension of its motion-direction. Three-dimensional objects in Euclidean space can also be thought of as four-dimensional objects where the fourth dimension has no length. When it was specified that all objects are moving in the w motion-direction at the speed of light, it was assumed, but not explicitly stated that the object's fourth dimension always aligns with the w motion-direction. Referring back to Figure_4.1, an object moving in the w-direction would be represented by a single point on the winding because its length in the w-direction is zero. Therefore, the object would not be hit by graveons. Consequently, the motion of the object would not be impeded if at rest or moving at a constant velocity. Suppose an object moves in the w-direction, but the object's fourth dimension does not align perfectly in the w-direction. Then, the object would have a finite length in its w-direction. Graveons would hit the object acting to impede its motion. This would cause the object's rear to move slower causing it to align perfectly. The object would provide for the fourth dimension, the same function that a trailing feather does for a dart. Note that the object does not have to rotate for the fourth dimension to align with the w-direction. Because the fourth dimension is zero length, it can be placed arbitrarily anywhere and align without the object having to rotate its position. It takes no force to make this alignment. This is the ultimate in streamline design.

OBJECTS IN POSITION-SPACE

Imagine a box with portholes on each of the six surfaces out in space. This box is moving at the speed of light in the w-direction. Since its fourth dimension has no length, portholes in the fourth dimension are not possible. Since the fourth dimension is aligned with the w-direction, none of the portholes faces the w-direction. Therefore, all the portholes face the other three dimensions, which is position-space. The automatic alignment of the fourth dimension of the object with the motion-direction, not only lets objects move at a constant unimpeded velocity, but also keeps objects facing position-space where it is not aware of its own motion and any other motion in its w-direction.

SPACE TRAVEL

The traveler's spacecraft moving at .8c from Earth to Soil, an Earth-like planet orbiting Alpha Centauri, 4 light-years away. The trip would take 5 years from the perspective of the rest frame, but 3 years from the perspective of the traveler. So, by traveling 4 light-years away, but clocking only 3 years, he has traveled faster than the speed of light, from that perspective. Nevertheless, at .8c, Earth and Soil appear 40 percent closer because of their difference in yaw with respect to the spacecraft. The spacecraft's instruments would still show the speed to be .8c, but since the distance appears 40 percent shorter, he expects to get there in 3 years. He does get there in three of his clock years, and presumably, he only aged 3 years.

Since his clock is running 40 percent slower than clocks on Earth and Soil, the trip really takes the same 5 years. This can be better understood with the aid of the following gyro-vector diagram that represents the spacecraft voyage from Earth to Soil, drawn approximately to scale. W_e represents the motion direction of Earth and v_s is the spacecraft velocity toward Soil. W_s is the constructed motion direction gyro-vector of the spacecraft. Gyro-vector v_e is the velocity of Earth relative to the spacecraft. It is equal in magnitude to v_s, but conspicuously not opposite to it, since the spacecraft velocity is so large, their position-space is vastly different. Nevertheless, unlike special relativity, observers everywhere would agree on when he left and when he arrived, after adjusting for the different clock-rates.

Figure 5.7 – Space Travel Gyro-Vector Diagram

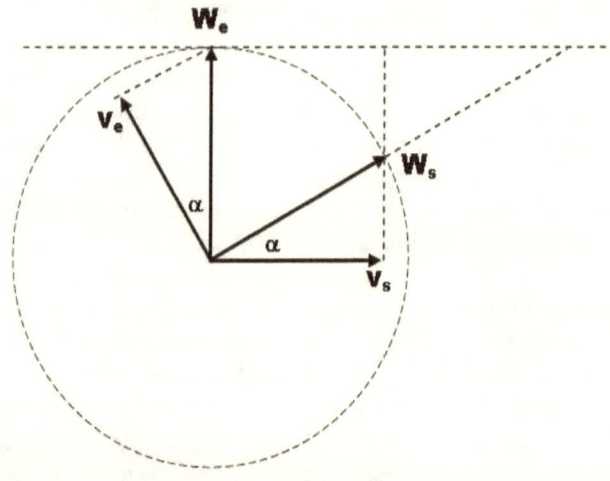

Figure_5.7 – This diagram, drawn approximately to scale, shows the gyro-vector diagram of the trip to Soil. Gyro-vectors W_e and v_s are Earth's motion-direction and the spacecraft velocity gyro-vectors respectively. The gyro vectors W_s and v_e, the spacecraft's motion-direction gyro-vector, and Earth's velocity gyro-vector, relative to the spacecraft, are constructed. The gamma factor is proportional to the arc sin α.

It is interesting how the gamma factor hides all differences of different inertial frames, contributing to a subterfuge. Distance shortening is due to the difference in yaw. Dilation of time and mass increase is due to the increase in inertial mass. The unvarying roll pitch product enable the relative velocity from both inertial frames' vantages to be the same. Finally, the exorbitant quantity of energy to effect this change is due to the large increase in force necessary to rotate the gyrovector further.

EHRENFEST PARADOX

This paradox was originally proposed by Paul Ehrenfest in 1909. He was born in Vienna in 1880. He made major contributions to the foundation of statistical mechanics, linking it with quantum mechanics. Some of his contributions were overlooked, because he worked at the University of St Petersburg, Russia, which was away from the mainstream physics community. However, he was appointed in 1912, to succeed Lorentz at the University of Leiden, Netherlands, where he stayed until his death in 1933. The paradox has gone through numerous machinations over the years, but essentially has not changed much.

Imagine a long train that stretches completely around a circular track, of radius R, where the first car is linked to the last car, similarly to all other car linkages. Also, visualize a train station platform stretching on the inside around the entire track. Now assume this train picks up speed, settling at a large fraction of the speed of light.

According to special relativity, an onlooker on the platform would measure the radius of the track to be R, independent of the motion of the train, since the track is stationary. In addition, the train is in contact, and riding on the track, its path must also be a circle of radius R. However, in accordance with special relativity, the onlooker notices that each of the cars gets smaller, and the space between the cars gets bigger, as the train picks up speed. Finally, the linkages are too short to hold the cars together, so they bust apart.

Now consider a passenger on the train carrying a Yard Stick, measuring all distances. He notices that the length, width and height of his car was the same at full speed as it was when stationary, consistent with relativity. In fact, as he walks between cars, he notices that the external width of each train car, and space between the cars have not changed. Consequently, his expectation is that the train remains intact, because all dimensions are exactly as it was before it started moving.

Notice the magnitude of the radius is not a factor in the analysis. It could be a small circle, or in principle, go clear around earth, all 25 thousand miles. However, as the radius gets larger, the centrifugal acceleration becomes less of a distraction, and the curvature less of a factor.

Only one of these scenarios can be true. Which is it? Does the train stay intact or does it come apart? Whichever happens must be true for both the passenger and the on-looker. That is the paradox. Einstein in 1915, the year he proposed general relativity, weighed in on this paradox, and explained it, only using special relativity, not general relativity arguments, which he completed that same year. For example, he expected distances in the direction of motion would shrink, consistent with gamma.

Subsequently, many others tackled the paradox but had to resort to more complex general relativity arguments, to try to resolve the paradox. It seems that whenever special relativity doesn't give the desired answer, the GR defense is invoked. That process is still ongoing.

The Gyroverse simplifies this analysis. In the Gyroverse theory, matter does not actually shrink. It only appears to shrink, because position space (and motion space) of objects in relative motion, are slightly oblique to each other. A small sliver of what is visible in one frame of motion is not visible in the other frame of motion. Nothing actually gets shorter, so nothing breaks apart. Furthermore, because the train is constrained to ride on the tracks, the two-dimensional plane containing the tracks remains part of the moving train's position space. Consequently, the train's length even appears to be the same for both the passenger and on-looker.

Nevertheless, position space of the moving train is still slightly altered, so something has to be different. If two of the dimensions of both the platform and moving train are forced to be the same, the third dimension must take up the difference. In particular, the height of each train car, as viewed by the on-looker, appears to be slightly smaller. In effect, the moving train is leaning slightly into the platform's motion space.

MUON DECAY

The famous Muon decay experiment is often used as indirect proof of special relativity's distance contraction, since there is no direct measurement of it. Muons from outer space, shower the earth uniformly. For this experiment, it was observed that the Muon flux at the top of Mt. Washington, the tallest mountain in north-eastern United States, was only slightly higher than the Muon flux at sea level. When considering the laboratory half-life of Muon, and time it takes for the Muon to travel the extra distance from the height of the mountain peak to sea level, it was expected that only a small fraction of the Muons should have been able to reach sea level.

According to special relativity, because the Muons strikes earth very close to the speed of light, time slows on the Muon, by the gamma factor. This stretches its half life enough to explain the much smaller than expected difference in the relative Muon flux between the mountain and sea level. On Earth's time does not slow, but the distance that the Muon must travel, according to special relativity, shrinks by the same gamma factor, also allowing the flux at sea level to be much larger than expected. This is cited as proof that speed shrinks distance.

The Gyroverse explains this phenomenon differently. For it, clocks run slower on the Muon, from both perspectives. In other words, the distance the Muon must travel does not shrink, but clocks, for example atomic reactions, simply run slower.

CHAPTER 6 - QUANTUM IMPLICATIONS

ENTANGLEMENT AND CLOSENESS

Quantum theory presents a dichotomy. From an analytical point of view, the theory is flawless. From an understanding of "how things work," it is extremely deficient. Of all the aspects of quantum theory, quantum entanglement underscores this dichotomy. Entangled particles are sister particles (usually photons or electrons) that share common origins and properties. According to the theory, even if the particles were separated to opposite sides of the universe, the particles would retain an immediate awareness of each other. While it has never been tested anywhere near that distance, it has been tested and verified to seven miles. Even for seven miles, this entangled particle influence on each other defies all logic. The process of understanding entanglement begins with the next figure.

Figure 6.1 - Representation of One Axis

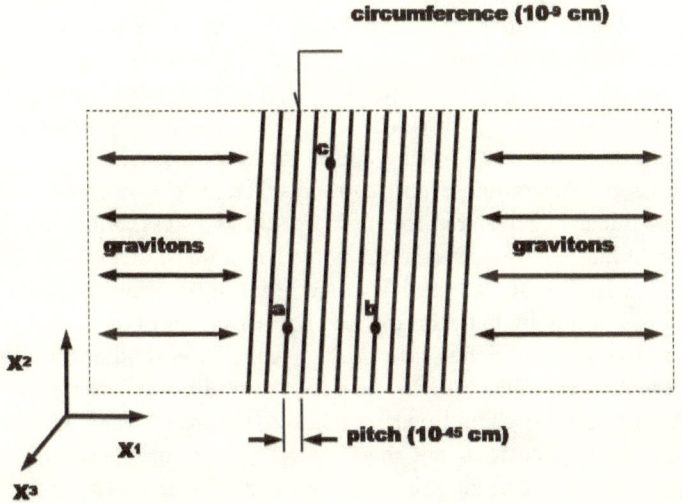

Figure_6.1 - The rectangular shape shown is a side view of a cylindrical shape, which represents the x-axis helix coiled around the hyper-cylinder. Particles a, b, and c are three particles on the x-axis. Particles a and b, on the x-axis, are at sister-locations. Particle c bears no special relationship with particles a and b. No loss of generality results in choosing the connection line between the particles as the x-axis, since the x-axis can be chosen arbitrarily. When a particle is on the x-axis its y and z coordinates are each zero.

Examining the three particles shown in Figure_6.1 illustrates that distances between points in twelve-dimensional space are almost not related to their distances in three-dimensional Euclidean space. Two random points further than atomic distances from each other in three-dimensional space would be of the order 10^{-9} centimeters from each other in twelve-dimensional space. Conversely, two particles seven miles apart might be as near as 10^{-30} centimeters from each other in the special case where the particles are exactly integral number of revolutions along the helix from each other. Such particles as a and b are at sister-locations. Particle c, on the other hand, is not at sister-locations of the other two. All distances in twelve-dimensional space are scaled down, so both these distances are small. Nonetheless, 10^{-9} centimeters is 21 orders of magnitude greater than 10^{-30} centimeters. If 10^{-30} centimeters were scaled to be the size of an atom, then an atom would stretch from Earth to the sun. When viewed this way, it puts into perspective the vast difference in magnitude between these two numbers. This is the situation when two quantum particles are entangled. They are much closer to each other in twelve-dimensional space than even two random particles within the same atomic nucleus. They are moving in unison as sisters, bound as one in this thin layer of separation in twelve-dimensional space. The force holding entangled particles together is approximately equal to the force of gravity between the two particles multiplied by the 10^{40} factor, the inverse of the slope of the gyrohelix. Note that the force decreases by the square of the distance. Although this force is small, even with this huge multiplier, it takes a significant force in three-dimensional space to break them apart. This results because they can move independently in Euclidean space, still remaining at sister-locations. Nevertheless, any apparatus that interacts with one of them is interacting with both. Particles that are not entangled might occasionally become entangled, but it typically can't be recognized. Viewing this situation in three-dimensional space, the universe would be made up of a grid of cubes about 10^{-9} centimeters on each edge. Figure_6.2 shows this grid, and a mapping of the same particles in Euclidean space. Particles a and b are in the same relative location in different cells. Small atomic particles can interact via these shortcuts in space. Feynman remarked, "Things on a small scale behave nothing like things on a large scale." As the distance between entangled particles gets larger, the sister-locations would eventually be separated by too much distance for the particles to remain bound, and they will completely separate.

Figure 6.2 - Three-Dimensional Space, Showing Universe Atom Boundaries

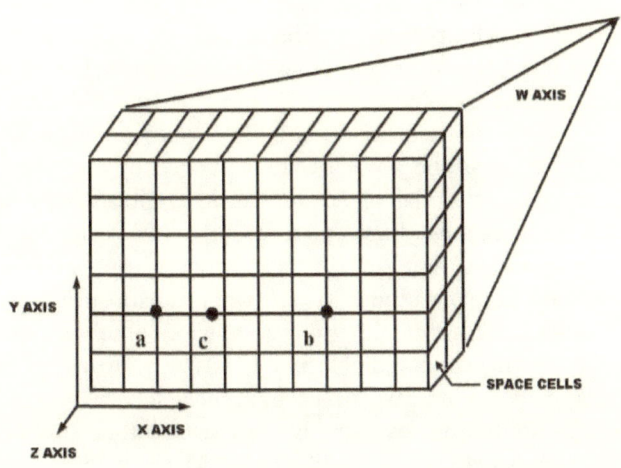

Figure_6.2 - This is a representation of three-dimensional position-space. Each cube edge of the imaginary grid, represents one complete winding of each axis helix. These imaginary cubes are called space cells. The points a, b, and c on both figures represent the three same particles in both views. Particles a and b are sister particles extraordinarily close to each other in twelve-dimensional space. They would be on the order of 10^{-30} cm apart, if they were seven miles apart in Euclidean space. Particle c is not a sister of either particles a or b. If electrons were situated at point a and b, they would have an immense attraction for each other, though they are far apart in three-dimensional space. This is the situation for entangled particles that are far from each other in Euclidean space, but very close in full space.

Given that an atom is 10^{-8} centimeters, and each cell is about 10^{-9} centimeters, it takes 10 revolutions, 10 cells, for each axis to enclose an atom. Thus, each atom spans about a 1000 cells. This entire figure is moving at the speed of light in the w-direction, the fourth dimension.

Even two entangled particles being as close as they are, the force pushing them together would be much to small to keep them together, if it were not for another factor. An atomic particle is not solid contiguous object, but a concentrated bundle of waves. Although these waves are dispersed over sister locations an imperceptible small distance, in three-dimensional space this small distance is magnified thirty-six orders of magnitude. For example, two particles that become entangled start out 10^{-45} cm from each other in twelve-dimensional space. Each particle being a tight bundle of waves, as the particles are separated by 10^{-30} cm (or 7 miles in three-dimensional space), the particles spread and coalesce around three locations. First, there are the two sister locations where the particles appear to be. Secondly, an intermediary common sister point continues to exist where a part of each particle is still located. This middle ground is where the two particle sections that remain exceedingly close, 10^{-45} cm apart, are being fastened together. The less of each particle left behind in this common location, the less attached to each other the particle become. At some point the bond breaks. The related action-at-a-distance phenomenon to entanglement is dispersement, the ability of a single atomic particle to be dispersed over large distances.

WAVE-PARTICLE DUALITY

Large objects must move from place to place taking the normal three-dimensional route in Euclidean space. Small objects, like electrons or photons, can take shortcuts by moving to sister-locations directly. The reason that large objects cannot take advantage of this shortcut is that all their atomic particles need to be aligned simultaneously with a dual set of sister-locations to keep the exact form of the object unaltered after the move. For an object made up of many atoms, this possibility is highly unlikely. A metaphor that might help understanding the situation is to think of the gyrohelix as a mesh screen. Particles smaller than the mesh opening can readily move unencumbered through the helix screening. Larger objects can't fit through the mesh and are forced to follow the path of the helix.

A photon is a tight wave packet that can remain small when viewed in twelve-dimensional space, but dispersed in three-dimensional space by spreading over sister-locations. When it strikes a large object, like a photographic plate, which is firmly in Euclidean space, it shows up as a single mark at one sister-location, since in reality it is a small indivisible single entity. Just in front of the photographic plate, the photon can be dispersed over many sister-locations, but when it strikes and marks the photographic plate that is massive and entirely in one three-dimensional place, it coalesces to a single dot. It takes only minute movements in twelve dimensions to show up in entirely different places in Euclidean space. Graveons, randomly striking a photon, probably provide this minute instantaneous random motion. A movement of 10^{-34} centimeters in twelve-dimensional space will translate to several meters in three-dimensional space. The concentrated bundle of waves in twelve-dimensional space is a wide-spreading wave pattern in Euclidean space. Its interference pattern determines where a photon can show up in three-dimensional space, hardly altering its location in twelve-dimensional space.

All light interference is caused by each photon interfering with itself. Coherent particles, for example, a laser beam, creates a narrower beam because the photon pack is traveling as a more massive tightly bound entangled object. Like any more massive particle, its wave pattern is more focused. Nevertheless, the individual photons remain separate, only interfering with themselves, not with other photons in the packet.

When the same photon is made to pass through two slits, the phenomenon of self-interference is more apparent. Even photons emitted one at a time will generate a diffraction pattern, but still make individual dots on a photographic plate. This shows that it can pass through both slits simultaneously, remaining as one object. If two different light sources were used to illuminate the different slits, no interference patterns would result, even if both were at the same frequency. This proves that photons interfere only with themselves. Electrons and nucleons are also wave packets, but experience interference less robustly than smaller photons. This quintessential double-slit experiment, illustrated in the following figure, epitomizes the strange aspect of quantum physics.

Thomas Young first did this classic experiment in the early 1800's. He was not only an English physicist, but started his career as a physician, with a primary interest in optics. Interestingly, he later became a notable Egyptologist, deciphering Egyptian hieroglyphics including the Rosetta stone. Feynman, very much associated with this experiment, pointed out that the entire mystery of quantum physics is embodied in this experiment. It is now sometimes called the Feynman double-slit experiment.

Figure 6.3 - Double-slit Experiment

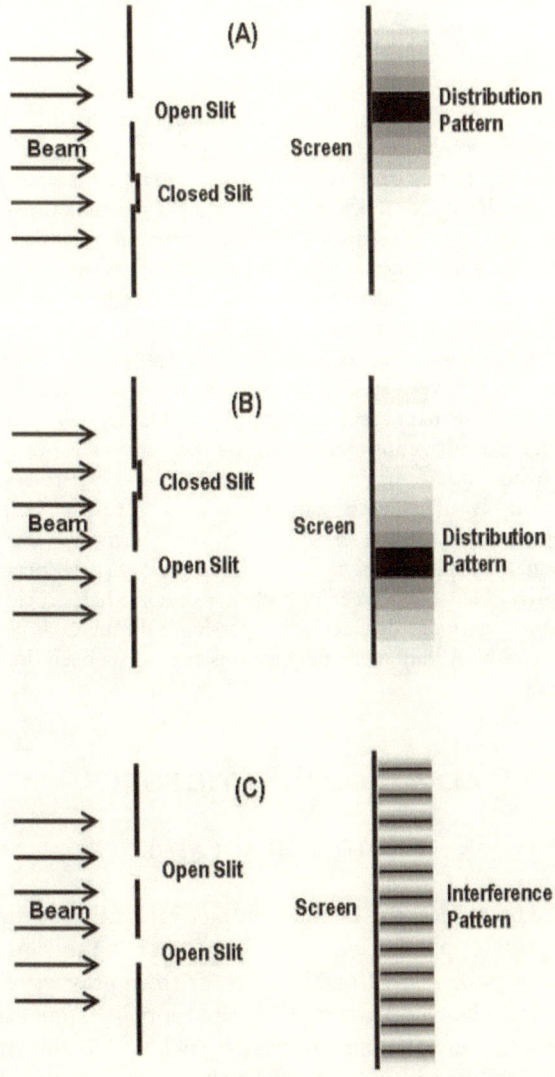

Figure_6.3 - This figure is made up of three illustrations. In the (A) diagram, on the left is a beam of photons. They pass through the top slit and hit the screen. The bottom slit is blocked. On the right is shown the screen pattern, rotated 90 degrees in order to be seen. Notice that just behind the open slit the screen pattern is blackest indicating that the photon intensity is greatest there. It gets lighter further from the slit. The (B) diagram is similar to the (A) diagram, except the pattern's greatest intensity is opposite the bottom slit that is open now. The bottom (C) illustration shows the screen pattern when both slits are unblocked. Here we get a diffraction pattern of alternate light and dark stripes and not a superposition of the (A) and (B) patterns. This is not surprising because waves are expected to interfere with each other. The separated slits would be expected to generate diffraction patterns of this type. What is most stunning is that when the photon rate is reduced to sending a single photon at a time, the same pattern results. In addition, if the pattern is reviewed very carefully, the dark areas consist of individual tiny black dots. On one hand, the diffraction pattern shows they are waves, but on the other hand, dots suggest they are individual particles. The explanation is that each photon passes through both slits simultaneously and interferes with itself. While orthodox theory has no reasonable explanation, in the gyroverse theory sister-locations in two separated slits are exceedingly close to each other. So close that the same particle is actually passing through both slits simultaneously. Additionally, this demonstration has been done with electrons.

ELECTROMAGNETIC FORCE

PHOTONS DUAL PATHS

Charged particles, for example, attract each other if the sign of their charge is opposite, and repel each other if the sign of their charge is the same. This force is different from gravity, which can only attract. The mechanism for repulsion is straightforward. Photons travel between like charged particles following the same trajectory used by moving matter. When the photon leaves a particle, the particle recoils, moving backwards. When it is captured by the similarly charged particle, the impact causes the second particle to move forward, away from the first. Photons jumping back and forth cause the particles to move away from each other.

While repulsion is straight forward, the interesting question is how both attraction and repulsion can be achieved. This can be understood with the help of Figure_6.4, starting at the top, where the helix represents the x-axis. Looking from the right side of the figure, the helix is wound in the clockwise direction. This is the path if a particle moves from one place to another and causes repulsion. Normally, photons also move along this path, but photons can take an alternate counterclockwise path, the horizontal mirror image of the clockwise path. The path is still pitched forward so that from a macroscopic view the photon is essentially following the same trajectory to its final destination. The bottom of Figure_6.4 shows the alternate path the photons can take going from one particle to the other. Both paths in Figure_6.4 are intertwined, and from a macroscopic point of view are overlapped, and appear as the same straight line. From an atomic viewpoint, photons taking the alternate path are passing atomic particles from the opposite direction. On the bottom path, when the photon leaves the atomic particle where it originates, the particle recoils propelling it forward. If it is captured by the target atomic particle, it propels the particle backward, propelling both particles toward each other, effecting attraction. While the photons can take both paths, the particle that the photons strike generally only takes the top clockwise path. From outward appearances, photons traveling the clockwise path, or the reverse path, appear identical. It is only at the atomic level where distances are too small to be seen that these differences exist. Most importantly, in a higher dimensional space, it is possible for the front and back of two objects to face each other simultaneously.

Figure 6.4 - The Two Photon Paths

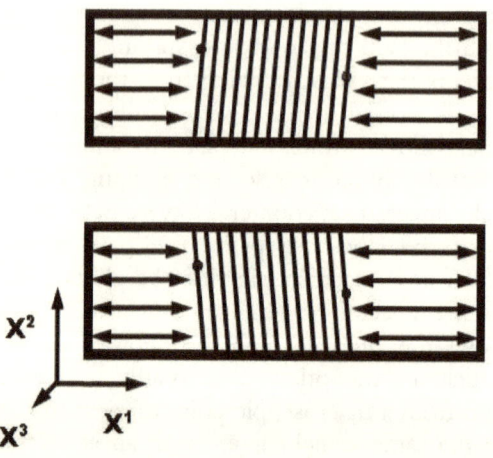

Figure_6.4 - This represents the two directions that photons can take in position-space. From a macroscopic point of view, both paths have the photons moving in the same x-direction. From a microscopic point of view, the bottom path has photons passing each atomic particle from the opposite direction. The top figure is the normal, clockwise direction of matter and photon movement in position-space. A photon from particle a to particle b causes a to recoil backward and b to move forwards, effecting repulsion between the two. The bottom figure represents the counterclockwise direction that photons can take to cause attraction. Notice that photons taking this path push the particles in the opposite direction. Since particles normally only move in the clockwise path on top, this photon motion causes attraction. Both paths intersect twice every turn of the helix, at a small fraction of an angstrom; so from a macroscopic point of view, both paths overlap.

DETERMINING DUAL PATH

An unanswered question persists. How do the photons distinguish whether they are headed for a positive or a negative charged particle, since they have to take opposite paths for each? A mechanism for photons to be able to decide automatically must exist. Advance notice or any handshaking between matter is not possible. When matter sends out a photon, it cannot know beforehand whether the photon is headed for a positive or negative charged target. Most likely, nucleons normally take the top path and electrons normally take the bottom path, so that when a photon leaves an electron and meets another electron it pushes it away, but if it meets a proton it pushes it toward itself. Likewise, electrons normally take the bottom path, so if a photon is released from a proton, it pushes other protons away and electrons toward itself. Antiparticles take the opposite paths. This mechanism ensures that like charges repel and opposite charges attract each other. Positrons take the same path as protons and thereby also attract electrons. A proton includes a positron, which provides the force. In effect, all electromagnetic forces are the result of photons emanating from electrons alone or as part of proton, taking the top or bottom path. When it takes the bottom path it is an electron, but when it takes the top path it is a positron.

Photons often take both paths. For example, a microwave oven or the sun can heat to a very high temperature, and yet cause little or no net force, because the water molecules are being pushed equally in all directions. In effect, current theory recognizes this by asserting that photons are their own anti-particle. Yet, photons emanating from magnets push each other greatly.

ANTIMATTER

Few phenomena stimulate the imagination as much as antimatter. It is the purest form of energy known, several hundred times more concentrated than the fusion that powers the sun.

Trekkies might remember Enterprise and Voyager's use of it for their powering. It is also ideal from an environmental point of view because it leaves no radioactive waste. Speculation persists that large regions of the universe might be made-up of it, although it cannot be detected. Others believe that it was all annihilated soon after the big bang. Still, it does not occur naturally anywhere near Earth, and cannot easily be made in quantity. It cannot be stored easily, because if it contacts the walls of the container, it will be annihilated with an immense release of energy. Nevertheless, containers, called "penning traps" using electric and magnetic fields, are being explored to store antiprotons. In addition, the technology for manufacturing it is being aggressively pursued. However, it not likely that anytime soon this technology will be viable, but some people continue to dream-on.

TRADITIONAL ANTIMATTER THEORY

One related phenomenon useful for exploring the dual paths and the electromagnetic attraction and repulsion mechanism is in the interactions of particles with their antiparticles. Each antiparticle corresponds to a different ordinary particle, such as a proton, neutron, or electron. They have most of the same characteristics of the particle, with the exception that they have opposite electrical charge and magnetic moment. Its most revealing unique characteristic is that when a particle and its antiparticle meet, they disintegrate each other into a burst of electromagnetic energy. Photons, being its own antiparticle, are thought to be an exception since its antiparticle has no distinguishing characteristic. The first antiparticle discovered was for the electron, and was predicted to exist by Paul Dirac, 1928, in his relativistic quantum theory. The relativistic label is applied because it is Lorentz transformation covariant, and consequently consistent with the requirement of special relativity. Dirac's theory explained the idea of particle spin and predicted the existence of positrons, electrons' antiparticle.

Another important physicist in the antiparticle discussion is Richard Feynman. He developed Feynman diagrams, a shorthand method for calculating and understanding interactions of electrons and photons. Each diagrammatic maps to the equation that governs its physical interaction. While the input and output of the process involves particles that could be detected, the process includes internal "virtual particles", its name stemming from the fact that cannot be detected, even in principle. These particles are short lived and violate the conservation of energy and momentum principle. Their justification is based on the uncertainty principle, which allows for conservation violations over short time intervals.

Almost all the common physical and chemical characteristics of matter can be attributed to these processes. Although these interactions also include positrons, Feynman defined positrons as electrons that travel backward in time. At first blush, a particle able to travel back in time could be presumed to exhibit all the requisite characteristics of an antiparticle.

To understand this, just imagine a motion picture of an electron run backward. This is about as simple a demonstration of time moving backward as can be envisioned. The electron would look the same. The rebound from collisions would also look much the same running backward. However, it would move in the opposite direction suggesting the opposite charge. In addition, clockwise rotation would appear as counterclockwise rotation, suggesting the opposite moment. This simplistic demonstration is compatible with the characteristics of the positron, which looks just like an electron, but with opposite charge and magnetic moment. If examined a bit closer, the illustration is misleading. For example, running backward, the mutual repulsion of two electrons would look like mutual attraction. Two electrons moving back in time are like two positrons moving forward in time, which should also be repulsive. In fact, if any ordinary scene were reviewed backward, it would immediately be recognized as such. It is only when the action is over simplified, one can be fooled.

Nevertheless, Dirac and Feynman had two different characterizations of antiparticles. Occam's rule would lead most dispassionate observers to conclude that accepting the positron as a distinctly new antiparticle is more believable than accepting that electrons can travel back in time. This position gets even more complicated because Feynman diagrams that give such accurate results, uses this characterization of backward time travel. For photons, an alternate explanation does not apply, since photons do not exhibit any different behavior, and are believed to be their own antiparticles. Once forced to accept that photons can travel back in time, it is no stretch of the imagination to believe that electrons can do the same.

Hence, two views on the nature of antiparticles persist. The Dirac view that antiparticles are distinct "mirror images" of elementary particles. This notion works well except when it comes to photons, whose antiparticle does not seem to have any distinguishing different characteristics from ordinary photons. On the other hand, Feynman suggested that antiparticles are just ordinary particles moving backward in time. This view is re-enforced by the observation that all the fundamental equations in physics seem to suggest that they hold equally well for negative time. Obviously, our experience in the macroscopic universe displays no evidence of time moving backward. In fact, in the macroscopic world, entropy is the standard argument that time only moves forward. This notion of entropy was originally formulated as the Second Law of Thermodynamics but has since been used more globally. One form of it posits that in a closed system, everything tends toward disorder, because the number of ways to create order is much less than the number of ways to create disorder. As a rule, in enough time, everything gets less useful.

GYROVERSE ANTIMATTER THEORY

In all deference to Feynman, when reviewing some of his statements on reverse time travel, one gets the feeling that his acceptance of particles moving back in time was without conviction, just waiting until a better explanation would be forthcoming. Despite the name antimatter, it is not strange that particles might exist of opposite charge and magnetic moment, since these are just ordinary characteristics of elementary particles. However, the fact that particles and antiparticles mutually annihilated each other is far from obvious. For two particles to cause such a violent interaction needs an underlining mechanism. Giving the particle a name that only suggests this violent reaction is not an adequate substitute for a real explanation. The gyroverse theory offers an actual explanation.

In the earlier section, it was shown how photons could travel dual microscopic paths for each identical macroscopic path. In other words, two straight-line intertwining paths between every two points exist. Clockwise is the most common path, the path of ordinary matter. As photons can travel in the counterclockwise direction, atomic particles can also travel in the counterclockwise direction. The difference is that while photons travel primarily in x, y, and z directions, atomic particles travel primarily in the w direction. When this alternate path occurs, atomic particles take on the characteristics of antimatter. It was already described how photons that move counterclockwise cause the opposite directional force on matter. Similarly, photons that travel the normal clockwise direction also produce the opposite force on antimatter, which travels the counterclockwise direction. This gives antimatter the characteristic of opposite charge and opposite magnetic moment. In addition, this same dual path mechanism will go a long way in removing the need for the "virtual" in virtual particles.

As previously discussed, the most important characteristic of antimatter is that when it meets ordinary matter, they mutually disintegrate in a burst of gamma ray energy. This occurs even if both the matter and antimatter are moving very slowly. Remember that matter at rest is moving in the w-direction at the speed of light. More precisely, it is rotating on the gyrohelix clockwise at the speed of light. Antimatter is also moving in the same w-direction at the speed of light, but it is rotating along a gyrohelix counterclockwise at the speed of light. Though from a macroscopic point of view, matter and antimatter might be moving slowly relative to each other, from a microscopic point of view they are rotating relative to each other at twice the speed of light. If they meet, a high-energy crash annihilates both resulting in an electromagnetic burst. In fact, because they have an opposite charge, they are even drawn to each other before the crash. Through this mechanism, all atomic particles can take on antimatter characteristics. Even a neutron with no net charge has its antiparticle that has no charge or magnetic moment but will annihilate each other on meeting.

Dirac solution for combining quantum theory and special relativity contained an electron with positive energy, and a positron with negative energy. He realized that relativity also had a negative energy solution that Einstein missed. Actually, that does not make sense, because if the positron really had negative energy, it should cancel both the electron and itself without producing a burst of energy. A positron has positive energy. It is just that it rotates on the gyrohelix in the opposite direction so that when it meets an electron, they attract each other crashing at twice the speed of light.

GRAVEON FORCE DERIVATIVES

In nature, seven forces are identified by this theory, all of which are caused directly or indirectly by the graveon. Included are four traditional forces: gravitational, electromagnetic, weak, and strong. Gravitational force is really two forces, the attractive force between matter as far away as neighboring galaxies, and a repulsive force for the universe as a whole, to be described later. Electromagnetic force also has two manifestations. For distances greater than a tenth of an angstrom, it can be either attractive or repulsive. For distances shorter than a tenth of an angstrom, it is only repulsive. Both manifestations of the electromagnetic forces are caused by the inertia of photons. However, photons get their inertia, as all other matter, from the force of graveons impinging on it. Finally, added to these forces are three other forces, all caused by the graveon. The first is the entanglement force that binds entangled particles to each other. The final two are the strong and the weak force that bind subatomic particles to each other.

THE STRONG FORCE

The strong force binds the atom's nuclei together. The strong force is about 10^{40} larger than the gravitational force. The weak force binds electrons (and antineutrinos) to protons to form neutrons. Its binding force is approximately 10^{32} larger than the gravitational force. As described in the gravity section, the gravitational force is on the order of 10^{36} times larger than its characteristic value in three-dimensional space, when acting in twelve-dimensional space. All of the atomic binding forces, both the strong force and the weak force are derived from the gravitational force. The force holding entangled particles together is also derived from the gravitational force.

At first blush, the gravitational force does not appear powerful enough to be the cause of the strong force, but on closer examination, it does have the correct magnitude. Gravitational force between matter is not between the atoms per se, but between the atomic particles that make up the atoms. In Figure_6.5, the nucleons, a, b, and c shown, appear as tiny specs, separated mostly by empty space. Two lines intersect nucleons a and b, the helix that is the x-axis, and the horizontal straight-line path that graveons take. Two atoms that bear no special relationship to each other, on average, have gravitational attraction in twelve-dimensional space proportional to 10^{36}. That does not mean that each nucleon attracts each other with exactly that much force. The reason two particles attract each other is that each of them blocks graveons headed for the other particle. It is this mutual blocking that creates a force imbalance, driving them together. Although nucleons a, b, and c are on the x-axis, only nucleons a and b are on the same graveon path. Some graveons passing a are blocked from reaching b, and likewise some are blocked by b from reaching a. Therefore, nucleons a and b will attract each other because of the reduction of graveons in-between them. Notice that nucleon c is on the same helix joining a and b, but is not on the graveon path intersecting a and b. It never blocks or is blocked from graveons striking a and b, and by that, does not develop a mutual attractive force with them. Because nucleons occupy one ten-thousandth the circumference of one turn of the x-axis helix, only 1 in 10^4 nucleons on the x-axis, are also on the same graveon paths. For every pair of nucleons that have a mutual attraction, 10^4 pairs have no attraction. This results in fewer nucleon pairs experiencing attraction, implying that the strong force is 10^{40} greater than the gravitational force not just 10^{36} greater. This is because in order for the average force between particles to be 10^{36} units, and considering that only 1 in 10^4 pairs attract each other, the ones that do attract each other, must do so with a force of 10^4 units. When nucleons are in neighboring turns of the helix, for example, sister-locations, they are just entangled. When they are on the same helix turn, they are atomically bonded.

This nucleon bonding of 10^4 units is just what is required to enable the gravitational force to be of sufficient strength to be eligible as an atomic nucleus binding force, the strong force. Note that if a nucleus fracture occurs, the nucleons will separate a small fraction of an atomic distance, but more than enough to move them relatively far apart in twelve-dimensional space. The binding force would then drop precipitously, although the particles are still only atomic distances from each other.

As an increasing number of nucleons bundle together, they increasingly find it more difficult to remain bundled. The electromagnetic force of repulsion grows faster than the attractive bonding force as the number of particles increases, limiting the maximum number of nucleons that can combine to form different atomic elements.

In summary, the theory suggests that gravity and the strong force are different manifestations of the same force. The atomic nuclei are held together in the longitudinal dimension by graveons permeating all space. This force, acting along the helix makes the gravitational force much smaller than the nuclei binding force, but effective at a much greater distance, leveraging strength for distance. It is 40 orders of magnitude weaker than the nuclei binding force, but acts 36 orders of magnitude further in distance.

Figure 6.5 - Graveon Derivative Forces

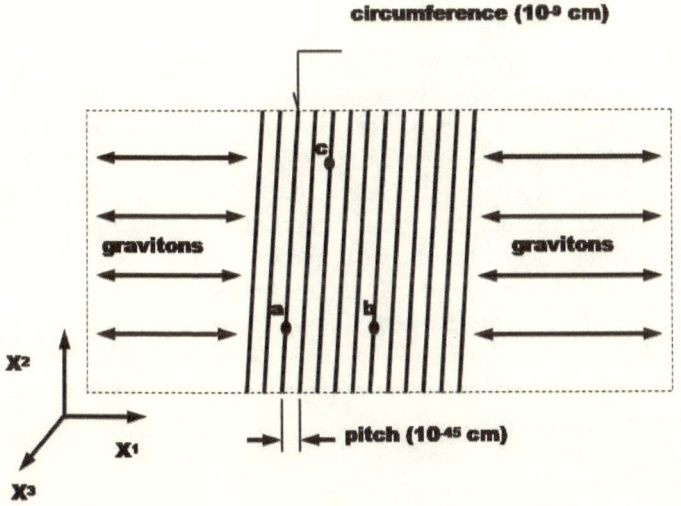

Figure_6.5 - These three nucleons all happen to be on the same helix, the x-axis. Nucleons a and b are on the same graveon path. Therefore, nucleons a and b are being pushed toward each other (causing attraction) because each stops some graveons headed for the other. Nucleon c is not on the same graveon path with the other two and does not generate an attractive force with them.

Because all three are on the x-axis, they are all at y = 0 and z = 0. Thus, nucleons a and b are in sister-location, but nucleon c is not. In effect, all locations on both the same straight line and same graveon path are sister-locations. Since the distance of one turn on the helix is about 10^{-9} centimeters and each nucleon is about 10^{-13} centimeters, about one in 10 thousand nucleons ($10^{-9}/10^{-13}$) are in sister-locations.

THE WEAK FORCE

As previously discussed, it is likely that this gravitational derivative force is the atomic nucleus binding forces, known as the strong force. Additionally, it is also the weak force, since it has approximately the right magnitude and characteristics. For nucleons, the force is roughly 10^{40} times the gravitational force and a very strong function of distance. The fact that the weak force is so much smaller results from the fact that an electron has only 1/2000 the mass of a proton. Since an electron is also much smaller than a proton, the force between them is smaller with a smaller range than the force between two protons. In twelve-dimensional space, two sister objects are flat up against each other. The smaller object would determine the force, since only the portion of the larger object abutting the small object would contribute to the binding force. Therefore, the weak force would not be 1/2000 the strong force, but about $(1/2000)^2$ times the strong force. Consequently, the gravitational force, the strong force, and the weak force and are one and the same. Decay associated with the weak force is caused by electrostatic force that is always repulsive at subatomic distances.

ENTANGLEMENT FORCE

The same force that binds the atom's nucleus together also binds entangled particles together. Because electrons are so much smaller than nucleons, 1/2000 the mass, the force between two electrons will be several million times less than between two nucleons, but still very large. Although this force is smaller than the repulsive electromagnetic force between electrons, which is 1/100 the magnitude of the strong force, it is acting over a shorter distance. An electromagnetic force only acts in three-dimensional space where the entangled particles are far from each other. The binding force, on the other hand, acts between the sister-locations in twelve-dimensional space where the electrons are very close together, making the binding force much greater than the electromagnetic repulsive force between electrons.

Entangled particles that are a few meters from each other in Euclidean space are only 10^{-34} centimeters from each other in twelve-dimensional space. Particles, that move seven miles from each other in Euclidean space, can move in unison in sister-locations, separating by less than 10^{-30} centimeters from each other.

The leverage between sister-locations in twelve-dimensional space and the corresponding Euclidean space is immense. It would take a large force to separate them in twelve-dimensional space, but because of leverage, about 36 orders of magnitude, only a tiny force to separate them in Euclidean space. Nonetheless, even while separating in Euclidean space they remain entangled, bound in twelve-dimensional space.

COMPOSITE PARTICLE MASS

A productive way to view the mass of an object is that it is proportional to the rate of graveon absorption by it. For example, if one object is absorbing graveon twice as fast as another, it has twice the mass. Consider the fusion of two protons, that is, the nuclear reaction of converting hydrogen into helium. Separately, the protons have their characteristic mass. As the protons are brought together, each blocks some graveons from striking the other proton. This results in a reduction in the graveon absorption by each, which leads to a decrease of the total mass when fused into helium. Consequently, the mass of helium is less than double each of the hydrogen atoms, before combining. The mass being smaller causes their energy mc^2 to be smaller. The missing energy results in the emission of electro-magnetic radiation, voila, the sun.

In general, larger atoms have less mass than the sum of their component parts. However, for very large atoms the packaging plays an important role. A large atom that happens not to give rise to as compact a nucleon arrangement as the component sub-packages, might have more mass than the component aggregate. Fissionable elements, like uranium-235 isotope, fall into this category.

This same process also applies to celestial objects. Consequently, the mass of two large objects decreases as the objects move closer together. For example, black holes have less mass than the amount of matter content would suggest.

FERMI & BOSE PARTICLE STATISTICS

One of the several ways of categorizing particles is whether they are "Fermi particles" or "Bose particles." The major attribute that distinguishes these particles is whether the particles can be bundled together. Fermi particles include all the elementary subatomic particles that we think of as particles, such as electrons, protons, neutrons, and some rarer particles. They are loners, keeping their distance from each other, obeying what is called "Fermi-Dirac statistics." This is especially evident from the "exclusion principle" where electrons in atoms in the same state keep their distance from the nucleus and from each other. A complete understanding of this is not apparent from orthodox considerations, but the result is that it is responsible for giving matter its characteristic structure. Instead of negatively charged electrons being tightly drawn to the positively charged nucleus as would otherwise be expected by applying Newtonian mechanic considerations, they keep a minimum distance. Empty space in atoms is 10^{15} times greater than the space occupied by the particles themselves. Matter, in some sense, consists mostly of empty space. The reason why Fermi particles keep their distance can be explained by the electromagnetic attraction-repulsion mechanism. Notice from Figure 6.4 that this mechanism requires that the particles be far enough apart for the photons to be able to take the reverse path to cause attraction. If the particles are close, both paths result in repulsion. The attraction mechanism can only kick-in at some minimum distance apart. When protons and electrons are very close, their electromagnetic force only repels them. As they get further apart, the attractive mechanism begins to function and prevents them from moving too far apart.

On the other hand, Bose particles, typified by photons, bundle together tightly, following the so-called Bose-Einstein Statistics. If n Bose particles are bundled together, the probability that another of the particles will join the bundle is proportional to n, the number of particles already there. Consequently, not only can Bose particles remain in bundles, but they also have an affinity for bundling. Laser light is a common example of this phenomenon. A complete understanding of why Bose particles behave this way is also not apparent from orthodox considerations. The gyroverse model can shed some light on this dilemma. The photons are being held together bound by graveons. If n photons were stacked this way, they would block n times more graveons, increasing their attractive force by a factor of n, which could explain the photon affinity for bundling. When the force increases, joining the bundle is easier for additional photons.

PARTICLE SPIN

One of the more disconcerting characteristics of atomic particles is spin. Quantum theory usually clarifies phenomena that were previously classically explained. Usually, quantum theory provides fine detail that classical explanations miss. The classical explanation becomes a simplified version of the real thing, leading to small errors. Spin on the other hand makes no sense classically and was dubbed "a pure quantum phenomena with no classical counterpart." The need for that admission alone should have been ample evidence that something was very wrong. However, before the gyroverse explanation of spin can be discussed, some background understanding of the composition of atoms is necessary.

HISTORICAL PERSPECTIVE

The atom can be crudely viewed as a miniaturization of the solar system where the atom's nucleus represents the sun and the electrons represent the planets. Planets are held in place by gravity pulling the planets toward the sun, counterbalanced by the centrifugal force developed by the planet's orbit around the sun. Likewise, the electrons are held to the nucleus by the electrostatic charge between the electrons and the nucleus; the nucleus is positively charged and the electrons are negatively charged. Their affinity for orbiting the nucleus encourages them to keep their distance. Unlike the planets, the electrons are all identical.

English physicist Ernest Rutherford proposed this earliest structure of the atom. He is considered the father of nuclear physics being awarded a Nobel Prize for it in 1908. The proton, neutron, alpha particle, and beta particle all owe their name to him. His model was problematic since orbiting electrons continually accelerate toward the center. Classical theory predicts that accelerating electrons would radiate, lose energy, ultimately crashing into the nucleus. That, of course, does not happen.

Niels Bohr, a protégé of Rutherford, armed with Rutherford's atomic model and the quantum ideas of Einstein and Planck, proposed that the electrons only revolve around the nucleus in a very limited set of circular orbits with discrete energy levels. He suggested that the outermost orbit determines the element's chemical properties. Although measuring the electron energy states of these orbits directly is not possible, every time an electron moves from a higher energy state to a lower one, it emits a photon whose energy is the difference between both energy states. An electron far from any nucleus has the most energy, and its energy decreases as it gets closer to the nucleus. By studying the emitted photon's energy spectrum, it is possible to determining all the energy states of the orbiting electrons.

Whereas planets could be any distance from the sun, and if slowed enough would crash into the sun, electrons cannot crash into the nucleus because an orbit of minimum distance from the nucleus exists. Electrons can only reside at discrete radii, denoted by a quantum number, n. It is closest when n = 1, four times further for n = 2, nine times further for n = 3, and so on.

ATOMS PHYSICAL STRUCTURE

While the Bohr Model described the key characteristics of the atom, it did not explain why only certain orbits were possible, and why the electron does not crash into the nucleus. The French physicist, Louis de Broglie, supplied a partial answer to this question. He developed a theory, built on the work of Albert Einstein and Max Planck, that matter also had wavelike characteristics. He developed the wavelength expression for a particle by analogy to the electromagnetic wave where ($\lambda = h/p$) and ($\nu = E/h$). For these equations, λ is the particle wavelength, ν

the wave frequency, E the energy, p the momentum, and h is the Planck constant. Since h is small, the wavelength of macroscopic matter is much smaller than the size of an atom, and never exhibits its wavelike properties. Alternately, if the de Broglie wavelength of the electron in a hydrogen atom is calculated, it is approximately the size of an atom. In fact, de Broglie used the wave idea to explain the orbits of the electron in the hydrogen atom. The electron orbit must correspond to an integral number of de Broglie wavelengths, to constitute a standing wave pattern that reinforces itself. This only occurs at the Bohr energy states. Nevertheless, it still does not explain why these stable states would not include the case where the electron falls into the nucleus. After all, the positively charged nucleus might be expected to attract the negatively charged electron. As previously explained, one new force of this treatise is that a positive charge will even repel a negative charge when the particles get too close to each other. Therefore, when the electron approaches the nucleus it is repelled, only allowing it to circulate at the Bohr orbits.

ADDITIONAL QUANTUM NUMBERS

Further precise measurements of the radiated frequencies showed the existence of more energy levels than the model predicted. These deficiencies were abated when it was realized that electron orbits did not have to be just circular, but could be elliptical or even absent. In addition, the orbiting electron charges would have magnetic moments that would also modify the radiated frequency. The number of quantum numbers was increased to three, with each combination defining a different state. In 1925, Wolfgang Pauli formulated the exclusion principle that no two electrons can occupy the same state simultaneously. Therefore, only a limited number of electrons can reside at each radius, as governed by Pauli's exclusion principle. Pauli received a Nobel Prize in physics in 1945 for this discovery.

Wolfgang Pauli was born in Vienna and was appointed in 1928 to a professorship of theoretical physics at the Federal Institute of Technology in Zurich. Being of Jewish decent, in 1940, he accepted a professorship of theoretical physics in United States, at the Institute for Advanced Study.

Simply stated, according to this postulate, two electrons in an atom cannot have all the same quantum numbers. When electrons get too close to each other, their mutual repulsion keeps them apart. Simultaneously, the nucleus attraction keeps them nearby. This fact is very important in understanding the chemical and physical structure of matter.

While the Bohr model was a major step in understanding quantum physics, it still missed the mark. First, it did not give insight why certain frequency emissions were more numerous. It also emphasized Rutherford planetary model that electrons were little charged balls circulating the nucleus. Finally, accurate measurements showed that more radiated frequencies existed than the model predicted.

Several different formulations were developed that explained this behavior. The first, developed by Werner Heisenberg, called matrix mechanics, was somewhat abstract. Heisenberg was born in Germany, received a professorship at the University of Leipzig, and later appointed director of the Kaiser Wilhelm Institute for Physics in Berlin. His most notable achievement was the Uncertainty Principle, a cornerstone of quantum physics. This posits that even in principle, knowing the simultaneous states of an atomic particle, for example, its position and momentum, is not possible. This idea shattered the determinism principle in physics, the idea that given all physical information one could predict the future outcome. He received the Nobel Prize in Physics 1932 as a founder of quantum mechanics.

Erwin Schrödinger proposed an easier, less abstract formulation of quantum mechanics, known as wave mechanics. It emphasized de Broglie's continuous wave nature of matter, as opposed to the matrix mechanics discrete jumps. It also rationalized the shortcomings of the planetary description of the atom. A big controversy ensued until 1926 when Schrödinger showed the equivalence of his wave mechanics and Heisenberg's matrix version of quantum mechanics. Soon after, Paul Dirac further unified both these formulations with his own theory.

At this point quantum mechanics could explain the behavior of the atom by its three quantum numbers, or so it thought.

Subsequently, more careful measurement of the energy spectrum of photons, as electrons fell from higher energy states to lower ones, showed that the frequency of each transition was duplex, signifying the presence of a fourth quantum number.

ELECTRON SPIN HISTORY

Up to this point, the discussion of the atom was limited to three quantum numbers, three of the four degrees of freedom attributed to the electrons within the atom. As already explained, these can be likened to the planetary orbits of the planets about the sun with a major difference that they are restricted to only certain discrete energy levels. These energy states result from the potential energy attributed to the distance that the electron is from the nucleus, the angular momentum of the electron as it orbits the nucleus, and finally the magnetic moment of the negatively charged orbiting electron. As already mentioned, these energies are not observed directly, but as electrons drop from higher to lower energy states, photons are emitted with the energy difference of the two states.

As mentioned, soon after these three numbers were discovered and explained, it was noticed that each state was not really a single frequency, but instead two different frequencies spaced very close to each other. These closely spaced doublet states are sometimes referred to as the fine structure, since the doublet frequencies are close to each other and hard to distinguish. The first explanation, which was wrong, but prevailed for several years, was that these dual energy states were caused by the relativistic effect of the orbiting electrons. The argument was that at the elliptic orbit's perihelion, closest point to the nucleus, the orbiting electrons move at a significant fraction of the speed of light, affecting its energy.

In 1925, two graduate students, George Uhlenbeck and Samuel Goudsmit, at Leiden, offered the present-day accepted explanation. They proposed that each electron spins on its own axis with an angular momentum of one-half the Planck constant and carries a magnetic moment defined as one Bohr magneton. Afterward, it was Paul Dirac, in 1928 that provided an elegant analytic solution, applying a relativity theory modification to Schrödinger's wave equation that described all the atomic particles' motion. He published this work in 1930 in a book, <u>Principles of Quantum Mechanics</u>.

Paul Dirac was born 1902, in Bristol, England and died in 1984. He held the Lucasian Chair of Mathematics, established in 1663, at Cambridge University. It is one of the most prestigious chairs, once held by Sir Isaac Newton, and now held by Stephen Hawking, a most prolific contemporary physics theorist and author. In 1933, he shared a Nobel Prize in Physics with Erwin Schrödinger for this work.

ELECTRON SPIN

The whole purpose of rehashing the previous quantum numbers of the atom was to put electron spin, the biggest mystery of the atomic characteristics, into a proper perspective. Theory has it that the electron has a spin with an angular momentum of $+h/2$ and $-h/2$. In other words, all electrons are spinning in one of two directions, for example, up or down. Because the electron is a charged particle, it also exhibits a magnetic moment corresponding to its spin and negative charge.

This characteristic of electron spin has some very troubling characteristics. It does not have any classical counterparts. It is tempting to envision it as a particle spinning around its own center with its surface having a negative uniformly distributed charge. This paradigm does not quite work. Given the magnetic moment of the electron and the maximum estimate for the size of an electron, the electron's outer surface would have to be moving more than two orders of magnitude faster than the speed of light to have that much magnetic moment. No one is willing to believe it could be moving that fast.

Interestingly, that argument is somewhat specious, since spin can't be used to send messages faster than the speed of light. In the case of quantum entanglement, which travels faster than the speed of light, the argument that it did not violate relativity rested on the presumption that entanglement could not be used to send messages. This revelation established the principle that going faster than the speed of light was not in violation of relativity, as long as it could not be used for sending messages. Apparently, this exemption does not apply to spin.

A most direct method for measuring spin would be by passing a stream of electrons through a magnetic field perpendicular to its motion. The expectation would be that electrons spinning clockwise would be deflected in one direction and those spinning counterclockwise would be deflected the other way. Unfortunately, a stream of electrons' wave properties would spread the wave packet, masking the effect of opposite spin. In addition, the moving stream of charged particles in a magnetic field also experiences force that further mask the effect of the spin. To get around this defect, two physicists, Otto Stern and Walter Gerlach proposed an alternate experiment, now remembered as the Stern-Gerlach Experiment. Instead of steaming electrons, they streamed silver atoms in a non-uniform magnetic field. A silver atom's outer most electron has no orbital angular momentum, so that the atom would not be deflected by orbital motion. In addition, the atoms are massive enough so as not experience any appreciable wave packet spreading. Finally, because each atom is electrically neutral, the stream through a magnetic field would not be expected to cause any deflection. Therefore, all the deflection that occurred was assumed to be due to the two orientations of electron spin that deflected the stream in two opposite directions, seemingly proving that the electrons were indeed spinning on their own axis.

Otto Stern later won a Nobel Prize in 1943 for also discovering proton spin, probably not as significant as his electron spin discovery. Perhaps, Gerlach did not also receive a Nobel because of his work on the atomic bomb, for Nazi Germany.

On the one hand, electrons exhibit a magnetic moment that is both verified by theory, and measurement. On the other hand, straightforward calculations imply that the amount of magnetic moment could only be caused by the outer surface of the electron spinning at two orders of magnitude faster than light. This seemed to create an insoluble situation. Nevertheless, physics has avoided this predicament by postulating that spin is not classical spinning, but that electrons all have "intrinsic angular momentum," a pure quantum characteristic with no classical counterpart. This is another instance of using catch phrases as a substitute for real understanding. To make matters worse, besides a magnetic moment, spin also has an angular momentum with no evidence of the energy that should result from it.

An old adage asserts, "If it looks like a duck and quacks like a duck, then it probably is a duck." As expected, another explanation for this dilemma exists. Remember that all atomic particles are circulating the gyrohelix in their respective motion-direction. This motion causes the particle's rest energy mc^2. Beyond that, a revolving charge would generate a magnetic moment. Because this spinning is not in position-space, but occurs in the hidden dimensions, it might be expected that only a fraction of its effect would bleed through. However, because the magnetic moment is almost an exact multiple of the Planck constant, it would suggest that the entire magnetic moment comes through, and none is hidden. Hence, by working backward, the radius of the gyrohelix could be determined more precisely. Assuming a 100 percent bleed-through, a good estimate for the radius of the gyrohelix can be calculated.

However, there is one more factor to consider. The angular momentum due to spin was taken as $+h/2$ and $-h/2$, even though it cant be measured. By so doing, the magnetic moment, which can be measured, is twice as much as expected. It was argued that spin is twice as effective as orbital motion in producing magnetic moment. Said concisely, the gyromagnetic ratio for the self-spinning electron is two times bigger than the value for an orbiting electron. However, since spin is also caused by orbital motion, albeit, in other dimensions, it should follow the same paradigm as the electron's orbital motion in an atom. Thus, the spin angular motion should also have been taken as $+h$ and $-h$, even though there is no possibility of it being zero.

Now, knowing that an electron is circulating the gyrohelix at the speed of light, with a magnetic moment of one Bohr magneton, the radius of the gyrohelix can be estimated using the following formula.

$$S = mcR = h/2\pi$$

Where: S is the spin angular momentum spin, m is the mass of an electron, c is the speed of light, $h/2\pi$ is the reduced the Planck constant and R is the radius of the gyrohelix.

Solving for R in MKS units:
$$R = h / 2\pi mc$$

$$R = 1.054 \times 10^{-34} / [9.11 \times 10^{-31}] \times [3 \times 10^{8}]$$

$$R = .0386 \times 10^{-11} \text{ meters}$$

Making the diameter of the gyrohelix:

$$D = 2R = .077 \times 10^{-11} \text{ meters} = .077 \times 10^{-9} \text{ cm}$$

The actual gyrohelix's diameter is about 25% of what is assumed herein (see Figure 4.1). In a certain sense, electron spin is the electron's primary quantum number, since it represents the atom's motion-direction energy, mc^2. Now that the gyrohelix radius is known one could go back and fine tune the appropriate equations, but that would not change the thrust of the treatise.

It turns out that 2π times the gyrohelix's radius, R, is equal to the Compton wavelength. Also, interestingly, the Bohr electron radius, the Compton wavelength, and the classical electron radius are each in the ratio of the fine structure constant (FSC) to the previous one.

$$FSC\ (\alpha) = e^2/2\varepsilon_0 hc = 1/137 \text{ (approximately)}$$

Arguably, the fine structure constant is the most important attribute of quantum physics. It characterizes the strength of the electromagnetic forces, correlated to the electron's charge. It is also roughly proportional to the anomalous magnetic moment. It's one of quantum physics greatest mysteries. discussed in Feynman's book QED.

Being able to sit on a chair without falling through, is due to the repulsion of the electric force between your body and the chair, carried by electrons. For example, neutrinos that don't respond to electric forces, have a mean free path in lead on the order of a light year. This illustrates that what we call a solid object is overwhelmingly empty space, made to feel solid by electric forces. Even standing on the ground without sinking in would be problematic without the electric force.

The absence of electric force's twin, magnetic force, would make motors and generators non-existent. All electric devices in the home would vanish. Nevertheless, mankind could survive because that's the way it was before Edison. However, most living cells contain a part called ATP Synthase, which is an electrical energy rotary motor engine. It causes among other thing muscle cells to contract, giving them their strength, without which they would not work, thereby making earth devoid of most living things.

The equation below is the spin angular motion for an electron. The Planck constant is derived from this equation. Below that is the classical electron spin magnetic moment. Notice both equations are similar where the electron's charge is substituted for its mass. But, the spin magnetic moment is only an approximation, because the location of the effective electron charge is not identical to its effective mass location. For this reason, the electron spin magnetic moment needs to be multiplied by a factor g, which is approximately .1% greater than one, even though Dirac's theory predicted it to be exactly one. (Note: in current theory g=2 because spin is taken to be 1/2)

The distance d from the center of mass to the center of charge is the ratio of the electron anomalous magnetic moment divided by the electron magnetic moment.

(1) Electron Spin Angular Moment $S = h/2\pi = \hbar = \mathbf{mcR}$

(2) Expected Electron Spin Magnetic Moment
$(e/m) \times S = \mathbf{ecR}$

The second equation is modified to:

(3) Electron Spin Magnetic Moment $g \times (e/m) \times S = \mathbf{gecR}$

where g is slightly greater than one

The Dirac equation predicted that g was equal to one. Because it was a surprise, that g was slightly larger than one, the part of the magnetic moment attributed to that difference was called the Anomalous Magnetic Moment. Significantly, it is proportional to the Fine Structure Constant. As previously mentioned, the Fine Structure Constant allows all living things to have energy, for examples muscles, and matter to be solid. What is most surprising is that arguably the most important universal constant is caused by "chance," whereby the effective charge of an electron is slightly misaligned with its effective mass.

Thus the anomalous magnetic moment, μ_{AMM}, of the electron to a first order is:

$$(g-1)eh/4m\pi = \mu_{AMM} \approx (\alpha/2\pi)(e/2m)$$

where α, is the fine-structure constant, e is the electron charge, m is the electron mass, h is the Planck constant.

The distance d from the center of mass to the center of charge is:

$$d = [(g-1)/g] \times \pi R^2/2\pi R = [(g-1)/g] \times R/2 = 2.238 \times 10^{-16}$$

which is about 8% of the size of the classical electron.

THE PLANCK CONSTANT

The Planck constant appears frequently in physics. As previously discussed, its primary appearance is as the spin angular momentum of the electron, and its magnetic counterpart spin magnetic moment. From the previous section, it is apparent that the Planck constant is one of the more basic physical attributes of the universe. It would be interesting to identify the connection of this primary appearance to all the others. Some of were discussed in the last section.

Another instance of h occurs in the energy formula hω. This occurs for a phenomenon known as Larmor precession. In this configuration electrons, for example, are inserted in a magnetic field. The electron's spin axis tends to align with or oppose the field. A potential energy is associated with these alignments. If it is opposed the potential energy is greater.

The precession frequency formula is:

$$\omega = 2\mu B/h$$

Where ω is the precession frequency, μ is the magnetic moment, B is the applied magnetic field, and h is the Planck constant.

When spin-flip occurs the change in potential energy E has the following form:

$$E = 2\mu B$$

Hence $\omega = 2\mu B/h = E/h$

This should be recognized as the form for electromagnetic energy. And indeed, when the flip occurs a photon with that energy and frequency is released, and when a photon of that specific frequency is captured by the electron it flips back.

DE BROGLIE WAVELEGNTH

As previously mentioned de Broglie, in his thesis proposed that matter would also display wave characteristics. That is to be expected since photons are just small particles dispatched very close to the speed of light c. Consequently, the de Broglie equation for and light and matter, for example an electron are the same:

De Broglie wavelength = $h/p = 2\pi m_0 cR/p = 2\pi m_0 cR/mv$

Where m_0 is the rest mass of an electron and m and v are the relativistic mass and velocity of the actual particle.

For a slow moving electrons m = m_0 so:

$$\text{De Broglie wavelength} = 2\pi Rc/v$$

However, photons are not just smaller pieces of matter because they cause electric and magnetic fields never reported using other fast moving particles.

ATOM FORMATION

For atoms to form, nucleons must clump together, and electrons need to stay apart, but still remain close to the nucleus. At the range of 10^{-15} m, the strong force, caused by the LeSage mechanism, is approximately 137 times stronger than electromagnetism, holding the nucleons together. At slightly larger distances, the strong force gets significantly weaker. Therefore, extra uncharged neutrons are needed to increase the binding force while decreasing the electromagnetic force, making it even easier for the nucleons to clump. However, electrons which are 1/1836 less massive than nucleons are dominated by the electromagnetic repulsive force separating them, since they have 25k times greater force than the force holding them together. In addition, the electromagnetic force between protons and electrons are repulsive when very close together, and only become attractive at about $.5 \times 10^{-10}$ m, keeping the electrons close, but away from the nucleus. The ratio of neutrons to protons generally increases with increasing atomic number, because the strong force weakens faster with distance up to a point.

STRING THEORY

String theory postulates many additional dimensions to unify all the forces of nature. The ten or eleven dimensions are defined, as the four prominent space-time dimensions, and the additional compactified dimensions, which are too small to be observed. In fact, the smallness was assumed just to explain why they were never observed. From previous discussions, clearly, the additional space generated by even tiny extra dimensions will far exceed the original three-dimensional space. Therefore, it is not clear why making them small would effectively hide them. The gyroverse model presents a more plausible explanation for the elusiveness of the additional dimensions.

There is another major difference between the two theories. String theory assumes the basic unit of matter is a stationery vibrating string, or perhaps a membrane. The basic unit of matter in the gyroverse model is a wave moving at the speed of light. Nevertheless, string theory and this model dovetail in that string theory has a strong mathematics foundation whereas this theory explains how the physical universe actually works. Perhaps some mathematical concepts in string theory could be applied to this model. Both have a comparable number of space dimensions, but arrive at it from a completely different thought process.

CHAPTER 7 - COSMOLOGY IMPLICATIONS

INTRODUCTION

Man has always pondered the heavens. For most recorded history, it was believed that the universe was geocentric; Earth was the center with the stars, sun, and moon put in place by the gods. Their purpose and operation were governed by fables that varied with each civilization. The Greeks, 2500 years ago, were the first in recorded history that developed an analytic model of the motion of the heavenly bodies, although it remained geocentric. Earth was the center and everything revolved around it. Five Hundred years ago it was determined that Earth revolved around the sun, known as the heliocentric model. Soon after, with emergence of the telescope, it was shown that the stars were not part of a celestial sphere, but actually celestial bodies equivalent to our sun. It was first proposed 250 years ago that the nebulae clusters might be other galaxies like the Milky Way. Just 75 years ago, these speculations were confirmed telescopically. To the present, the size estimates of the universe steadily increased, but what is more important, it has become more accurate. The observable universe is 13.5 billion light-years; enabling man to peer back to the universe's supposed infancy. Several techniques have been developed for estimating the distance different celestial bodies are from Earth. These use parallax shift, standard candles, and Doppler shift, each of which is discussed in the Cosmology Overview chapter.

The universe consists of 10 billion galaxies, averaging about 100 billion stars each. Galaxies are further grouped into clusters with up to thousands of galaxies each. Some clusters are grouped further into super-clusters. Over small regions, the universe does not appear homogeneous. The space between galaxies and clusters of galaxies is vast as compared with the size of a single galaxy. However, when partitioned into regions encompassing many clusters each, the universe does seem homogeneous, having the same density of matter throughout. In addition, the universe seems isotropic; that is, it looks the same in all directions. By some estimates, accepted by the mainstream physics community, the universe has much more dark matter than visible matter. One of several clues to this prediction comes from observing the rotation of spiral galaxies. The rotation causes each star to have an outward centrifugal force acting to break the galaxies apart. Conversely, the gravitational attraction between the celestial matter provides the opposite centripetal inward force holding the galaxies together. This type of effect can be experienced if a heavy object tied to a string is swung. This outward force is felt, but if the string were let go, the heavy object would fly away. It is necessary to provide the pulling force to keep the object revolving. Calculations reveal that the visible matter cannot nearly provide sufficient gravitational force to hold the galaxies together. It is speculated that unseen "dark" matter, constituting about 90% of the galaxy's mass, provides the additional binding force. Another, more plausible, explanation for holding the galaxies together, which does not require dark matter, will be discussed further along in this chapter.

COSMOLOGICAL MODEL

At the time Einstein developed general relativity, it was believed that the universe was static. The solution of these equations for the universe included an unknown constant that by default was initially set to zero. This solution required that the universe be either expanding or contracting. To be consistent with the paradigm of the day that the universe was static, neither contracting nor expanding, Einstein subsequently introduced a positive cosmological constant to exactly counterbalance the contracting force of gravity. Edwin Hubble later found out that the light from distant galaxies experienced a red shift, highly suggestive that the galaxies were receding from each other, contradicting the static universe hypothesis. Einstein subsequently backtracked and removed the cosmological constant, calling its inclusion "the biggest blunder in my life." Later, it was concluded that the cosmological constant lead to a very fragile solution that was unstable. Any slight imbalance would grow to a large imbalance and predestine the universe to collapse or expand forever. The slightest expansion would intensify the repulsion force and reduce the contraction force; the slightest contraction would cause the opposite outcome. In effect, general relativity only had two stable solutions, where the cosmological constant was zero, shown in the following illustration.

Figure 7.1 - Three Universe Expansion Scenarios

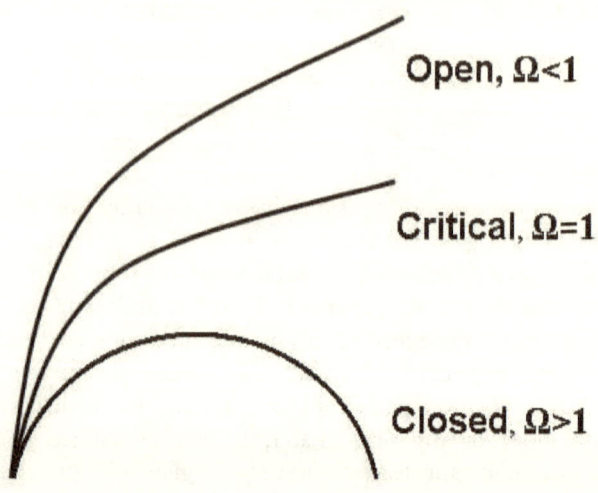

Figure_7.1 - These are three expansion scenarios. The top one is one that approaches a final constant expansion velocity. The bottom case shows the situation where the expansion does not approach the critical point. Consequently, gravity reverses the expansion causing contraction and an eventual collapse in a big crunch. In the center critical case, the expansion speed approaches zero, but does not reach it. However, this case assumes perfection, and from practical considerations does not exist. An infinitesimal variation will result in one of the first two cases.

From this point on, the predominant thought was that since the universe was expanding, it must have been much smaller earlier. This led to the hypothesis that the universe was originally concentrated at a single point that underwent an expansion that continues to the present. Working backward from the present speed of expansion, the big bang was estimated to have occurred 13 to 15 billion years ago. This hypothesis, in short order, led to two problems, the flatness problem and the horizon problem.

THE FLATNESS AND HORIZON PROBLEMS

Flatness Problem

The flatness problem surfaced because calculations indicate that for the universe to have evolved as it did, the force of expansion at the big bang had to be balanced almost perfectly, to better than one part in a trillion, with the gravitational attraction, during its initial expansion. To achieve this perfection, matter must be uniformly distributed in every direction, with a mass density and expansion rate perfectly balanced. The imbalance will grow geometrically for each doubling of the size of the universe. Considering the universe has doubled in size over 150 times since inception, even a small starting imbalance is greatly magnified. This would have been a very unlikely accident of nature. If the expansion were any slower, the universe would have collapsed long ago. If it were faster, the density of matter would be so small that it would never have clumped together to form stars and galaxies.

This problem gets exasperated because the original mainstream cosmology theory presumes that the expansion force was short lived, only present while the universe was still microscopic. Thus, any initial imbalance could not be corrected over time. Of the 150 doublings of the universe that occurred during its 14 billion year age, over two thirds of them occurred in the first day, while the last one took 7.5 billion years. After the first day, any appreciable initial imbalance would preordain an unworkable universe. It is this front-loading of the universe doublings that makes it understandable how only a slight imbalance can be so detrimental. Citing another example, when a rocket is fired, the major thrust occurs immediately, but it is the corrective subsequent smaller thrusts while in flight that enables the rocket to go into a precise orbit.

The measure used to determine the flatness is defined by the parameter, omega (Ω), the ratio of the total quantity of matter in the universe to the critical amount. This is labeled in the previous figure. The critical amount of matter occurs when the universe experiences neither unlimited expansion nor eventual contraction with Ω equaling exactly one. If it is greater than one, then the universe will eventually contract. On the other hand, a number less than one means the universe will expand limitless forever. Best estimates today are that Ω is between 0.1 and 2.0. Working backward, this implies that Ω had to be almost exactly one to within less than one part in a trillion initially. Conversely, according to orthodox theory, if the expansion parameter, Ω, initially were not balanced perfectly with the gravitational attraction, the universe would not be as flat as it is today.

Horizon Problem

The horizon problem became evident when the cosmic background radiation, referred to as CMB, coming from space in all directions, was measured and found almost identically the same. The most precise of these observations were made by the COBE satellite, which stands for Cosmic Background Explorer, launched in 1989, with a follow-on WMAP satellite in 2001. The measurements implied that matter at the outer reaches of the universe, in all directions, must have been homogeneous earlier, for the radiation to be so uniform.

This cosmic microwave background radiation is a remnant of the early years of the universe, when it was 300 thousand years old; the time after which radiation was uncoupled from matter and able to travel unimpeded in the universe. Just before that time all matter in the universe consisted of atomic nuclei and free electrons because the universe was too hot for the particles to combine into neutral atoms. Charged particles are very efficient at scattering photons, not letting them stray far from where they were generated. Nevertheless, due to the temperature of matter, they continually emitted their very distinctive frequency spectrum of electromagnetic waves, called black body radiation. The distribution of frequencies for this radiation only depends on the temperature, not the material makeup. Once the temperature of the universe fell below 3000°K, neutral atoms formed. These atoms became relatively nonreactive with photons, and the photons began their unencumbered journey in space, sometimes referred to as the "decoupling epoch" or "afterglow." Because the universe expanded by a factor of a 1000 since then, the remnant blackbody radiation comes to us with the same distinctive recognizable frequency spectrum, but at one-thousandth the frequency as when the radiation originated.

The uniformity of background radiation seemed problematic, in that the origin of the radiation were never thought to have been in causal contact with each other, making it unusual for them to have produced the same amount of radiation. For example, the universe has been observed to distances of 13.5 billion light-years in each direction, indicating that light would have taken at least 27 billion years to travel from one end to the other. Applying conventional thinking, these places could not have been in contact since the universe is thought to be only between 13 and 15 billion years old.

Inflation Theory

This led to a dilemma. The recession of all the galaxies was a strong indication that the universe had a big bang origin, but the flatness and horizon problems were inconsistent with that presumption. Consequently, another theory called the inflation model was proposed, which had the entire universe expanding uniformly many times faster than normal in its infancy. In 1981 Alan Guth, with further important contributions by Andrei Linde, Paul Steinhardt, and Andy Albrecht developed this theory. It relies on some well-known quantum physics principles, quantum fluctuation. While it is a bit of a stretch to apply a quantum phenomenon to the whole universe, Guth recognized that General Relativity has an implied relationship between Ω and the geometry of the universe. Inflation would make the geometry flat and would drive Ω extremely close to unity. In less than a second, the entire universe grew from less than the size of an atom to something greater than today's visible universe. Accordingly, our present-day visible universe was generated from only a tiny region of this inflated beginning.

A two-dimensional analogy to inflation is the over-worked confetti on a balloon metaphor, where the confetti separates as the balloon is inflated. If the balloon size were doubled, the distance between the confetti would also double. The inflation model resolved the flatness and horizon problems, and is the theory widely accepted today. This theory allowed the universe to grow uniformly, increasing the separation of matter, without matter having to move in the conventional sense. Unfortunately, the inflation model introduces a third theory to supplement quantum and relativity theories. One of the remaining troubling aspects of this theory was that certain globular clusters were estimated to be older than the universe. In addition, the expansion and distribution of matter in the universe would have been so uniform that it is problematic how galaxies could have ever formed.

THE GYROVERSE BIG-PUSH MODEL

Introduction

While the big bang model is widely accepted, it had inconsistencies with subsequent astronomical observations, and needed several modifications. Most recently, some data suggests that not only is the universe expanding, but its expansion is accelerating. This is inconsistent with expectations, since it was assumed that gravity would work to slow the expansion, even if it were not great enough to stop it. This evidence will be presented, but further on, it will be shown that gravity isn't slowing the universe expansion, but it is not accelerating it either.

A reasonable big bang theory should explain how the universe started as a subatomic spec and mushroomed into the gigantic proportions of our existing universe. The correct mechanism should not need any auxiliary mechanism to explain the horizon or flatness dilemmas. It needs an expansion method that caused the universe to expand uniformly. Finally, an attraction mechanism is required that explains the formation of the stars and cluster of stars. The gyroverse model has all these attributes, without twisting or augmenting the theory.

The Gyroverse
The Hidden Structure of the Universe

In review, the gyroverse is constructed in twelve dimensions. The three Euclidean dimensions of position-space are actually composed of three unique dimensions each, rolled up in a tightly wound helix, around a torus. Therefore, nine of the twelve dimensions are used to construct position-space, the x, y, and z three-dimensional space that is familiar. The fourth or motion-direction, where all matter is moving at the speed of light, is identical to the other three, but is hidden from observation. Since most matter close to Earth is moving in relation to each other very slowly, they share almost the same three position-space dimensions. In the gyroverse, three perspectives for understanding and visualizing physics must be considered. For most ordinary motion, three-dimensional position-space is all that need be considered. In contrast, for the more complex interesting questions, the higher dimensional perspective must be considered. At the next deeper level, or using the metaphor, another peel of the onion, the universe has another dimension, hidden, but constructed like the other three. All matter is moving in this fourth dimension, in a straight line, at the speed of light. In the three-dimensional or four-dimensional view, the universe appears Euclidean and very large, but is finite. With another peel, the universe is recognized as being wrapped into a tight construct composed of twelve dimensions. Each of the Euclidean axes is actually a separate three-dimensional tightly wound helix. The motion-direction is special since all matter is circulating this helix at the speed of light, creating an immense gyroscopic action, causing the gyrohelix. In this view the universe is tiny, only an angstrom wide in each direction. Everything in the universe is extremely close to everything else. Nevertheless, to get from any place to any other place the gyro action forces matter to take a circuitous path that is over 36 orders of magnitude greater than the shortest distance. As described previously, the universe has very little intrinsic mass, at least 28 orders of magnitude less than what is observed.

Possible Creation Scenario

From a twelve-dimensional perspective, the universe started with an r/p ratio of about what it is today, but with a zero r-p product. In other words, the amplification factor was its present value, but the intrinsic energy and mass were zero. The subatomic particles had a uniformly distributed yaw. From a four-dimensional perspective, all matter in the universe was initially contained in a tiny four-dimensional hyper-box. Since the universe was hot, atoms had not formed, and all matter just consisted of free atomic particles. In the Model Basics chapter it was described how all of the atoms in the universe, including empty space could fit into a four dimensional box of atomic proportions. Since atoms had not formed yet, the box could even be four or five orders of magnitude smaller, the size of the largest atomic particles. In principle, all these particles could fit into this small space without being compressed, initially having very little kinetic energy.

The mere presence of matter caused high speed emission of graveons in all directions resulting in the universe expanding during this initial period. The mass of matter was less, and clocks ran faster. In a sense, this period in the universes' evolution took longer in local time than it appears in retrospect. Other matter within the aggregate subsequently captured these graveons. This action produced an outward force that accelerated the expansion, increasing the roll-pitch product. For the first 300 thousand years when the universe was still above 3000 degrees and filled with just atomic particles and neutral atoms had not yet formed, communication between particles was not just confined to moving along the helix. Subatomic particles were entangled and able to communicate more freely across the helix, spanning vast distances almost instantly. Thus, electromagnetic radiation was not needed for homogenizing all matter.

Universe Expansion

Once the particle spacing in the initial four-dimensional subatomic size box expanded, it formed a four-dimensional hyper-sphere where all matter was on the outer surface or hyper-shell. The hyper-shell, or 3-sphere is three-dimensional, where every particle is moving outward at the same speed uniformly expanding the hyper-sphere. Matter is also uniformly distributed on the hyper-shell. Initially this three-dimensional hyper-shell subset space contained only individual subatomic particles, not yet combined to form atoms. Each atomic particle is still nominally the same distance from the big bang origin, expanding uniformly outward in the four-dimensional hyper-sphere, increasing the size of the hyper-shell. These particles first formed into neutral atoms as the universe cooled and then formed into the large celestial bodies. By this time, all matter moved nominally close to the current roll-pitch product, at the speed of light, but at different motion-directions, such that the fastest relative velocity between any objects is twice the speed of light. The universe hyper-shell is closed; circumscribing it, returns to the same location. This reflects the geometry of the hyper-shell and is not dependent on whether the expansion is accelerating or decelerating. However, all the universe energy is fueled by the universe expansion energy. Without expansion there would be no energy in the universe. Also, because gravity is orthogonal to the expansion direction it is not an important factor in slowing the expansion. Any future potential collapse will be chaotic, not nearly as precise as the original expansion, and probably will not cause another cycle of an orderly universe creation. The following diagram will help in understanding this universe creation process.

Figure 7.2 - The Universe Model

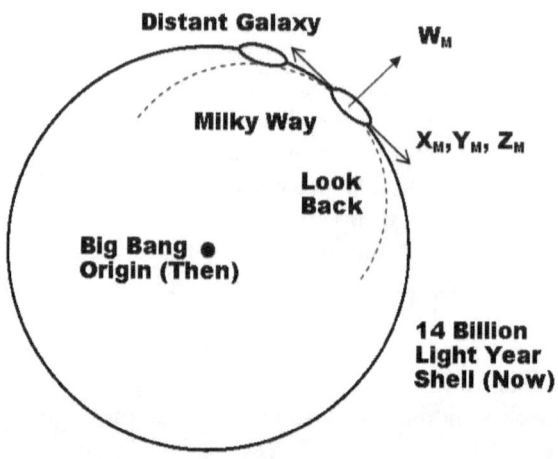

Figure 7.2 - This figure represents the four-dimensional view of the entire universe. The 14-billion light-year circular thin hyper-shell, or 3-sphere, represents our three-dimensional view containing all the matter in the universe. The subatomic size spot at the hyper-shell center is the location of the big bang that occurred some 14 billion years ago. From this tiny atomic size beginning, where everything in the universe fit comfortably within, the universe ballooned at the speed of light to its present size. The interior and the exterior of the extremely thin circular hyper-shell are completely void of photons and matter. The disk on the surface of the hyper-shell at two o'clock contains all the stars in our own Milky Way galaxy. The nearby disk at one o'clock is a distant galaxy. Every other galaxy in the universe is also on this hyper-shell. The galaxies are not to scale, appearing orders of magnitude larger than their actual size. The dashed logarithmic spiral curve segment represents looking back in time at celestial objects from Earth. Besides being distant, they are also seen as they were when the universe was younger and contained in a smaller hyper-shell.

The four space dimensions are illustrated at the Milky Way location. The line tangent to the circle represents the three position axes, x_m, y_m, and z_m. Light is constrained to travel in local position-space as it traverses the universe. This makes the universe appear Euclidean and uniformly distributed, although the light reaching Earth originated from the universe when it was smaller. Gravitational force keeps photons and matter contained within the expanding hyper-shell. Everything we experience comes to us tangential to our location along the universe hyper-shell. Instead of the universe appearing as a three-dimensional global hyper-shell, it appears as a three-dimensional flat hyper-sheet. The entire Milky Way, held together by mutual gravitational force, is moving in unison in the radial W_m-direction at the speed of light, so that the hyper-sphere is actually a hyper-polygon. Because everything around us is also moving at the same speed and in the same direction, we are not aware of this motion. For example, the rest energy of matter defined by the equation $E=mc^2$ is not really rest energy, but the kinetic energy of all matter due to the expansion velocity in their respective w-direction. The distant galaxy shown is moving in a slightly different radial direction. The difference in direction causes the relative speed between these two galaxies. As the universe hyper-shell grows, like confetti on an inflated balloon, everything moves away from everything else proportionally to their distance apart, giving the appearance of expanding space.

For every graveon captured by matter, another graveon is released. This repeated action generates a large force, causing the matter to expand with considerable force at first. As the density of matter decreased, fewer of the graveons were captured because each graveon met less matter, reducing this expansion force. Initially this expansion force caused the expansion velocity to accelerate rapidly. The expansion was not one event of short duration, but occurred over an extended period. Hence, what is called the big bang would be better characterized as the big push. Although the expansion force was reduced, as the universe got larger, it was not eliminated, but now is probably too small to be easily detected or have any appreciable effect. The recent revelation of an accelerated expansion is probably due to something else, since in this epoch the expansion mechanism described here would be too weak. Whereas the expansion force is primarily in position-space, it has a small component in the w-direction where the expansion takes place. While the universe as a whole was expanding, gravity (LeSage mechanism) caused attraction of local matter, forming the stars and clusters of stars. The force that produces attraction of local matter is primarily acting in position-space, unlike expansion that predominately occurs in the motion-direction. Because galaxies are wide, opposite ends would tend to move in slightly different w-direction. This would create an expanding force tending to drive the galaxy apart. Gradually, gravity tended to bind the entire galaxy, encouraging it to move in the same motion-direction. Once this occurs, it no longer drifts apart. The overall expansion of the universe in the motion-direction keeps matter uniformly distributed around the 3-sphere, but locally clumped into stars and galaxies. Each mechanism hardly influences the other.

It takes much less force to move objects in their present motion-direction because it does not result in gyrohelix precession, changing its yaw. Without a change in yaw to contend with, the force needed to move celestial objects are reduced by about 28 orders of magnitude. At this force reduction, it would be possible for an average-man to move a mass as large as earth, if he had a place to stand. Thus, the expansion force is very efficient in charging-up the gyrohelix with kinetic energy.

While from our perspective conservation of energy applies for most common situations, the universe as a whole circumvents this. It did not take much energy to produce this universe, because it took place in the motion direction. When energy is retrieved, much more is obtained than it took to produce it, because it happens in position space, forcing the gyrohelix to precess. It is interesting that conservation of energy, a pillar of physics, is not as universal as assumed.

This expansion resembles but is not entirely equivalent to the current accepted paradigm, where the universe expansion is the stretching of space, and local motion is traditional inertial motion. Note that the gyroverse mechanism for the universe expansion is different, in that it is real movement of matter, although non-inertial, since the force is directly in the motion-direction. The expansion velocity can continue to enlarge the universe, while gravity holds the galaxies and cluster together. Therefore, gravitational attraction between matter in the universe would not appreciably slow the expansion. Only real movement of matter enlarges the universe. Matter initially was not very energetic, moving at a small r-p product as compared with present times. The pitch being small appears to make the universe larger, since it takes many gyrohelix revolutions to transverse even short distances between the objects. This small pitch also increases the size of large objects, as well as the measuring standards. Consequently, the relative size of objects are not affected by the change in pitch. It is like viewing an enlarged copy of a map. Everything on the map increases in size proportionally, including the scale used to measure distances on the map.

However, from the perspective of atomic particles, they were smaller before the expansion. This difference between atomic particles and atoms is interesting. If the pitch would double, it would take half as many revolutions to traverse an atom. On the other hand, because an atomic particle's size is only a fraction of a revolution, it takes the same fraction of a revolution to traverse the atomic particle independent of the pitch. This is why the size of large objects does not change, but the size of atomic particles is larger now.

PHOTONS AND LOOK BACK

Photons are also affected by the expansion. The photons experience the same expansion force as matter and move with matter, hugging the hyper-shell. When a distant galaxy (or bright star) is viewed, the light emanating from it, takes a logarithmic spiral path, approaching along the hyper-shell in position-space, where Earth is now, and not from the direction where the light began its journey toward us. Since the growth of the universe is in the w-direction, any asymmetry of the universe is in the w-direction, not in position-space where it would be apparent. Thus, the universe appears flat and not spherical. In position-space, the universe just looks like everything is separating uniformly from everything else in that space. The next diagram illustrates this point.

Figure 7.3 - Virtual Location of Objects

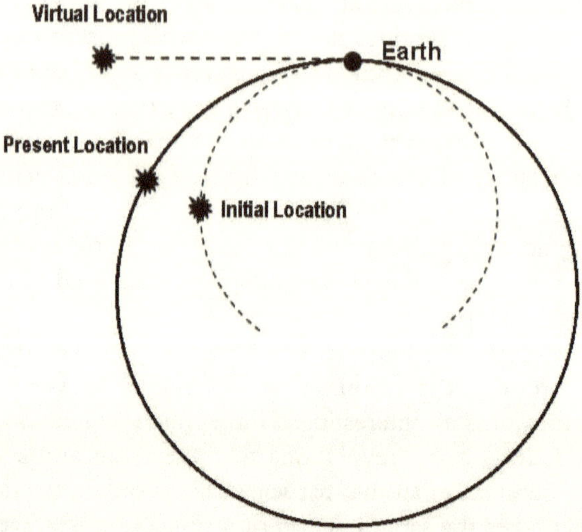

Figure 7.3 - This figure is a diagram of the universe hyper-shell showing Earth on top. When light leaves a celestial object, it does not travel in a straight line toward the present location of Earth, but instead spirals toward Earth, holding close to the hyper-shell as it grows. The look-back dashed line spiral intersects the initial star's location where light was emitted several billion years ago when the hyper-shell was smaller. To the left is the present location of that star. Starlight, bent gradually by the gravitational pull toward the hyper-shell, arrives on Earth tangent to it. Now it appears to be coming from the virtual location shown, which is in Earth's position-space, and a distance intermediate between its present and initial location. This makes the universe appear Euclidean.

Two aspects of the photon's participation in the expansion merit further discussion. Photons from distant stars were emitted tangential to the hyper-shell from the star's past location. At that emission angle, the photon might soon find itself outside the hyper-shell. Since all matter is on the hyper-shell, the photon would be gravitationally attracted toward the hyper-shell tracking the expanding hyper-shell until it reaches Earth. In fact, any matter straying from the hyper-shell would be attracted back to it. During our entire observable past, the expansion speed has not changed much, nor has the speed of light. In other words, lightspeed from all objects close or distant is the same.

On the other hand, the wavelength of light is increased proportional to the growth in the hyper-shell from emission to capture. To illustrate this point, assume that a photon starts one unit long, with no tension between the front and rear of the photon. As the hyper-shell grows, the front and back of the photon each expands radially, in direct proportion to the relative growth. This is equivalent to its wavelength growing proportionally to the expansion. Thus, the speed of light from distant stars is the same as the present speed of light, while the wavelength of light from a distant star is increased proportional to the growth of the hyper-shell.

UNIFORM EXPANSION

The continuous expansion force causes the universe to grow more uniformly than the single outward thrust assumed with the orthodox big bang model. This can be appreciated by considering the entire universe, and some very large spherical subset, as viewed from a three-dimensional perspective. It is dominated by the graveon expansion mechanism, and not gravitational attraction. Uniformly distributed matter will continue to expand uniformly, with any slight imbalance being self-correcting. This will ensure that on a large scale the universe will grow uniformly.

To understand this mechanism, assume that matter in the universe is spread throughout space uniformly. If the large sphere is considered, the matter in that sphere is proportional to the space in the sphere, or the cube of the sphere's radius. The quantity of graveons emanating from that matter is also proportional to the amount of matter. Since the matter is spread uniformly, the quantity of graveons escaping from this large sphere is proportional to the surface area of the sphere, or the square of its radius.

Therefore, the graveons' outward force and acceleration of matter from the sphere are proportional to the amount of matter in the sphere divided by surface area of the sphere. This is proportional to the cube of the radius divided by the square of the radius, which reduces to being simply proportional to the radius. If the acceleration were proportional to the radius, the radial velocity of all matter would also be proportional to the radius. Consequently, if matter were originally uniformly distributed, it will remain uniformly distributed after some more expansion. Every object will be moving at a velocity relative to every other object proportional to their distance apart. This expansion is achieved not by matter moving in position-space, but by the expanding universe hyper-shell. The universe looks as if it is inflating, but in reality, everything is moving away from everything else by conventional motion, albeit noninertial, in their respective motion-direction. The motion-direction of objects close to each other get locked together, and move in the same motion direction. Objects that are far apart move at the same speed, but in far different motion-directions having a large relative speed difference. Said differently, everything is moving at the same roll and pitch, but with a uniformly distributed yaw.

No process is perfect. If the distribution of matter were sparser for larger radial distances, the outward force would grow slower than the radius of the sphere. This would reduce the acceleration at large radial distances, retarding the expansion and making the distribution of matter more uniform. Conversely, higher density at larger radial distances does the opposite, thereby increasing the outward acceleration and velocity lowering the density of matter at larger distances. Where there is too much matter the growth is accelerated; where there is too little matter the growth is retarded. Notice that this entire process is not contingent on how much matter is contained in the universe for it to be driven to flatness. It does not depend on matter getting a very precise initial shove or its density having any particular value.

SIZE, LOOK AND ENERGY

Since the age of the universe is 14 billion years, and the hyper-shell is expanding at the speed of light, the radius of the universe hyper-shell is roughly 14 billion light-years. Even if the actual age of the universe were older than 14 billion years, because it was expanding slower at first, it would not affect the radius estimate. It would have been expanding slower by just the right compensatory amount to make using the present Hubble age in the computation correct. For example, if the universe were always expanding at its present-day expansion rate, it would have expanded to 14 billion light-years. If it expanded at half the speed, it would have taken twice as long, but it still would have expanded to 14 billion light-years. In other words, using the current value of the Hubble constant provides a better estimates for the size of the universe than it's age. Consequently, the full length around the universe today is about 88 billion light-years.

The universe is much smaller than current orthodox theory assumes. At the present maximum look-back distance of about 13.5 light-years, just about all of the matter in the universe is observable, when taking into account that the universe was small when the light was emitted. A galaxy that is 13.5 light-years away is about halfway around the 3-sphere, at the threshold of being visible from every direction. Interestingly, two celestial objects 13.5 billion light-years away from us in opposite directions, are very close together, the other way around, and are in causal contact. The Euclidean appearance of the universe is a mirage.

There are important differences when the expansion mechanism is compared to conventional motion. All conventional motion occurs in the three dimensions of position-space. The resulting motion of an object alters its yaw, caused by a slight precession of the gyrohelix and an accompanying increase in r/p ratio. This gyroscopic rotation magnifies the force required to cause this motion. The graveon expansion mechanism is different, since it occurs in the fourth dimension of space. The universe expanded, by increasing its speed in the motion-direction directly. It does this just by increasing its r-p product, without a simultaneous change of yaw; thereby not by precessing the gyrohelix. The force that is needed to give it the added velocity is infinitesimal as compared with what would be normally required. This expansion mechanism also enables the relative speed of objects to be greater than the speed of light. Actually, the current expansion speed becomes the new speed of light. The speed of light limitation of objects, comes about because conventional motion requires that the gyrohelix precession always accompany an increase in r/p ratio. The force required to accelerate an object is proportional to the product of the r/p ratio with the change in yaw. This product builds to infinity as the desired velocity approaches the current speed of light. Being able to increase velocity without precession finesses this limitation. All these factors reinforce the illusion that the universe is being inflated rather than just moving in the conventional sense. Appearances can be deceiving.

MATTER FORMATION

While the expansion mechanism works well in a uniform distribution of matter, the LeSage mechanism works poorly in it. The reason is that all matter is pulled uniformly in all direction, going nowhere. However, a local excess of matter overwhelms the smaller opposing force and pulls more matter to it. Doing so, the imbalance becomes greater and additional matter is pulled toward it. Unlike the repelling force, which is stable, driving toward uniformity, and correcting matter's imbalance, the attractive force is unstable, magnifying any non-uniformity. From an engineering perspective, expansion entails negative feedback, and gravity has a positive feedback mechanism. Though in small regions slight imbalances cause matter to lump, over large regions the expansion remains uniform. Over time, slight imbalances in matter distribution allowed stars and galaxies to form. Over many galaxies, matter in the universe remained uniformly distributed, not having had sufficient time to lump together. As galaxies moved further and faster away from each other their window of opportunity for the forming of galaxies is passing.

FREQUENCY SHIFT IN STAR RADIATION

Two paradigms for explaining the red shift in the light emitted by celestial objects must be considered. The more classical explanation is Doppler shift. Doppler shift is experienced when a police car with its siren blaring passes a person. First, the pitch is high as it approaches, and then lowers as it passes. In this paradigm, the difference in speeds between the transmitter and the receiver causes the Doppler shift.

The second paradigm is the general expansion shift of the universe. In this paradigm, the velocity of light also remains constant, but the frequency shift is proportional to the universe expansion ratio. The expansion ratio is indicative of the comparative size of the universe when the light was emitted and received. Neither of these paradigms measures distance directly. Each, in principle, leads to a different red shift. Doppler shift only depends on the relative velocity of the celestial object when the light was emitted and Earth's velocity when the light was received. During the intervening time, it does not matter what the relative velocities were. On the other hand, the frequency shift due to a general expansion of the universe does not depend on the relative speed of Earth and the celestial object when the light was transmitted or received. It only depends on the relative sizes of the universe over that entire period. Consequently, both calculations will usually lead to different results, but since the expansion was constant and uniform, both paradigms lead to similar results.

Overall, it is correct to assume that the uniform component of expansion is attributed to the expansion shift, and the local variation of each object's motion be ascribed to Doppler shift. For very distant objects, the expansion shift governs the calculation. For close objects, the Doppler paradigm dominates.

AGE OF THE UNIVERSE

Because the expansion has been accelerating, the big bang must have begun earlier, making the universe older than previous estimates. The smaller the entire universe, the larger the force of expansion, since it is more likely that graveons emanating from matter will strike other matter sooner. This likelihood is proportional to the inverse of the radius of the universe squared or the force of expansion is $F_a = k/R^2$. This force was large at first, but much smaller now. Nevertheless, it has continuously increased the universe's expansion speed. The Hubble formula for the receding velocity between two celestial objects is:

$$v = Hd$$

Where v is the receding velocity of the two objects, H is the Hubble constant, and d is the distance between the two objects.

This formula expresses the velocity that distant objects move away from Earth. A first estimate for the age of the universe is obtained by observing that if the current receding velocity was caused by the big bang, then objects moving apart were original in contact v/d time ago. This ratio is recognized as the inverse of the Hubble constant or 1/H. In other words, 1/H is a first estimate of the Hubble age of the universe. However, based on the prevailing wisdom that gravity slows the expansion, and dark energy has worked to accelerate it, the accepted age of the universe was refined to 13.7 billion years old.

This was accomplished using the WMAP satellite, launched recently, which detailed the structure of the CMB, cosmic microwave background radiation. It accurately determined the cosmological constant and other key parameters.

However, in the gyroverse model, the gravitational force is orthogonal to the expansion, so that gravity would have little effect on the expansion, making the universe somewhat older, and consequently has been rounded up to 14 billion years. Conversely, the expansion does not interfere with the gravitational force consolidating matter into stars and galaxies, suggesting that they were formed much earlier. Gravity and the expansion are not working against each other, resolving the paradox that certain celestial bodies appear older than the universe.

UNIVERSE PSEUDO-ACCELERATION

As aforementioned, recent measurements suggest that the universe is not only expanding, but its expansion is accelerating. This is more consistent with the gyroverse model theory, but surprising for the orthodox inflation theory, where it was assumed that gravity slowed the expansion. The cosmology community has explained this new revelation by adding a cosmology constant back into general relativity equation that Einstein himself added and then discarded many years before. This time, it was needed to explain an accelerated expansion of the universe, and not to counterbalance gravitational contraction. Just as the original constant was troubling to Einstein, the current constant is troubling because it not a constant of nature, but is an adhoc way to explain the accelerated expansion. To explain this force, it has to be assumed that empty space is not entirely empty, but contains a huge amount of "vacuum" energy.

Two independent groups, in 1988, simultaneously came to this conclusion. The first group, Supernova Cosmology Project was lead by Saul Perlmutter, a member of Lawrence Berkeley National Laboratory's Physics Division. The second group, the high-z supernova search team was led by Robert Kirshner and centered at the Harvard University Center of Astrophysics, where he was a professor. Both groups had many participants working on various aspects of the problems and their work was chronicled in several books including Kirshner own book <u>The Extravagant Universe.</u>

Combining two different distance measurements on a Type 1a supernova led to this acceleration conviction. These type supernovas are certain white dwarfs near the end of their life that go through a swift huge nuclear reaction, producing immense predictable peak luminosity. Their light output is so bright that they can be seen almost back to the origin of the universe. If the peak light output is measured on one of them in a distant galaxy, by using the inverse square law on it, the distance of its parent galaxy can be deduced. Alternately, the parent galaxy distance can be indirectly measured using the more standard red shift. When this dual measurement procedure is applied to a Type 1a supernova, the luminosity distance measurement indicates that the galaxy is further from Earth than the expansion shift measurement suggests. For remote galaxies, the distance difference becomes progressively greater. The red shift measurement, which is really measuring the receding velocity of the galaxy, being less than expected, infers that the galaxy's receding velocity was less in erstwhile eras. This implies that the galaxy's receding velocity has been accelerating. This acceleration caused gamma to increase, clock time to slow, and the speed of light to increase.

The gyroverse model includes a clear mechanism that causes attraction force within galactic distances and an expansion force for the universe as a whole. A slight expansion force persists, implying that the universe expansion is still accelerating, but it is much too tiny to be of consequence in explaining the recent measurements.

While the aforementioned observations does suggest that the expansion is accelerating, a study of Figure 7.4, would illustrate that this comes about, because of a misunderstand of the physical structure of the universe.

Figure 7.4 - Virtual & Actual Path of Light

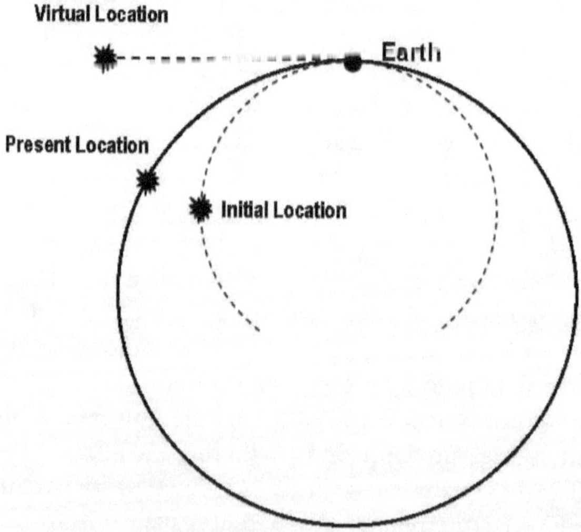

Figure 7.4 – The current paradigm assumes that starlight comes to us along a straight-line path from the virtual location (believed to be the actual location). However, starlight actually comes to us along a spiral path segment from the stars real initial location. The curved path is longer, making the starlight dimmer than expected. Currently, this phenomena is explained by assuming the expansion is accelerating, which is also consistent with this observation, but not correct.

To appreciate this point of the apparent accelerated expansion, refer to Figure 7.4. Notice that light from a distant galaxy does not follow a straight line, but rides the expanding circumference of the 3-sphere, following a spiral path. For galaxies close to the Milky Way, the circumference and the straight-line chord distance are almost the same. However, for far-off galaxies, the distance along the curved path is much longer, and accordingly reduces the amount of light received, making it appear that the universe expansion is accelerating. This contrasts with the current expansion space paradigm, which assumes a straight-line light path, as evident by examining the assumed path from the distant galaxies' virtual location to the Milky Way. In the current paradigm the virtual location is believed to be the actual location. But what is especially important is that the virtual location makes the universe appear Euclidean.

COSMOLOGY POTPOURRI

DARK MATTER HYPOTHESIS

In 1933, the astronomer Fritz Zwicky, born and educated in Switzerland, studied the motions of distant galaxies. Zwicky came to the United States in 1925, where he became a professor of astronomy at Caltech. From their rotation, he could estimate the mass of a group of galaxies. He also estimated the total mass of the group by measuring their brightness. It seemed that the mass obtained by studying their rotation inferred that they were two orders of magnitude greater than their brightness estimate. While his estimates were crude, subsequent studies essentially confirmed his results. This discrepancy in the brightness and rotation mass estimates spawned the "missing mass problem."

This observation lay dormant until the 1970's, when scientists began to realize that only large amounts of hidden mass, referred to as "dark matter," could explain this discrepancy. Dark matter gets its name because it does not absorb, reflect, or emit electromagnetic waves, the signature that matter is present in a particular region of the universe. Originally, the most common forms of baryonic matter, containing protons, neutrons, and electrons did not qualify, because they absorb and emit electromagnetic waves. The class of dark matter is now expanded to include any matter that cannot easily be observed.

One class of candidates for dark matter goes by the acronym WIMPS (weakly interactive massive particle). These exotic non-baryonic particles have not been detected but are theorized to be 10 to 100 times more massive than protons. Some examples are axions, heavy-neutrinos, and photinos. Another candidate for dark matter is neutrinos. Initially it was thought that neutrinos were massless, but recent evidence suggests that they may have a small amount of mass. Their huge abundance and weak interaction with baryonic matter make them an attractive candidate. A third candidate is MACHOS, (massive compact halo objects). These objects include black holes, matter so dense that nothing, not even light can escape from it; orphan planets that escape and no longer orbit their solar system; brown dwarf stars that spent all their nuclear fuel; and dwarf stars without enough mass to ignite. None of these candidates has been detected in adequate abundance to explain all the missing matter.

Overall, galaxies rotate appearing to be neither expanding nor contracting. The centrifugal force caused by the galactic rotation works to pull the galaxy apart. On the other hand, the gravitational pull on all the matter in the galaxy tends to draw it together. When these two forces are in balance, the galaxy maintains its size. The gravitational attraction decreases in proportion to the inverse of the square of the distance the star is from the center of the galaxy. The centrifugal force decreases just inversely proportional to its distance from the center. To keep these two forces in balance the speed of the stars must also be slower at greater distances to further reduce and match the inward force.

By observing the difference of the Doppler shift of stars moving in the plane of rotation of a galaxy, the velocity of stars can be estimated. This velocity, together with the star's distance from the center is sufficient to estimate the centrifugal outward force of the stars. Since the gravitational force counterbalances this force, the quantity and location of matter in the galaxy needed to produce this force could be calculated.

An important person in this cosmological saga was Vera Cooper Rubin. She was born in Philadelphia, Pennsylvania in 1928. She graduated from Vassar College in 1948 with a bachelor's degree in Astronomy, completing her doctorate at Georgetown University in 1954. In 1970, with very precise examination of the Doppler frequencies at the outer limits of galaxies, she discovered that stars further from the galactic center were not moving slower. Instead, stars at the outer limits reach a minimum speed after which the speed does not decrease further. This meant that the outward force of these stars was much greater than originally thought. To offset this larger force, much more matter in each galaxy was needed, increasing the evanescent dark matter estimate to about 7.5 times the visible matter. This result was startling because it removed any hope that all the missing matter could be collected from less exotic sources. In addition, it was surprising that almost all the missing matter had to be placed far from the galactic center, and not more uniformly distributed.

The dark matter hypothesis is perplexing. It is hard to understand how the universe can be mostly composed of this material with little traces of it on Earth. Still, the hypothesis is congruous with contemporary physics, as well as the gyroverse theory. An alternate theory with a substantial following is even more compatible with the gyroverse theory. It is not accepted by the mainstream, because it is incompatible with relativity. This theory, discussed in the next section, goes by the acronym MOND, short for Modified Newtonian Dynamics.

Donald Wortzman

MODIFIED NEWTONIAN DYNAMICS

Whereas the dark matter hypothesis is the current predominant paradigm, MOND (Modified Newtonian Dynamics), an alternate theory, has significant support. This theory was first proposed in 1983 by Mordehai (Moti) Milgrom, Professor in the Department of Condensed Matter Physics at the Weizmann Institute in Israel. Moti challenged the idea that dark matter was needed far from the galactic center to provide enough gravitational attraction to hold the galaxy in place. He proposed the alternate possibility that the inertial force might be smaller than Newtonian (or general relativity) theory predicts for slowly accelerating masses. If $F=ma^2/a_0$ for accelerations below a_0 with the Newtonian $F=ma$ only applying to accelerations greater than a_0, he could explain why the stars very far from the galactic center could move so fast, without flying away from the galaxy. For example, if the radius of a star to the galactic center is doubled, the gravitational force is reduced by a factor of four. When the velocity remains the same, as happens for stars far from the galactic center, the acceleration only halves, reducing the outward force by only a factor of two. Then the outward force overpowers the force of gravity, tearing the galaxy apart. If force were proportional to the square of acceleration, instead of just simply proportional to it, the outward force would also decrease by a factor of four, counterbalancing exactly the gravitational force. This could explain the galactic formation and containment without resorting to the dark matter assumption. Moti estimated a_0, the transitional acceleration to be about $1.2*10^{-8}$ cm/sec^2, a very minute rate of acceleration. At that acceleration, it would take roughly 75,000 years for an object starting from rest to build up to the speed of a jet aircraft. At the acceleration a_0, the transition point between both formulas, the centrifugal force for both have the same numerical value.

It is likely that the transition occurs gradually and not sharply as described. Another way to view the phenomenon, more consistent with the Gyroverse theory, is that the inertial formula doesn't change, but the inertial mass is what is changing. Inertial mass is not an intrinsic quality of matter. When an object accelerates, it causes the gyrohelix's plane of rotation to precess. It is this occurrence that gives matter its mass-like quality. In this paradigm, the inertial mass of an object would be zero until a force is applied, causing it to rise asymptotically to its characteristic inertial mass value. It is difficult to measure this on earth, because its rotation gives all matter a large background acceleration.

The MOND theory has very compelling characteristics and standing alone would probably have much wider acceptance. The thought that more than 95% of the matter in the universe cannot be observed is discomforting; MOND gets around that. Unfortunately, MOND is inconsistent with general relativity in that it contradicts the equivalence principle, the underpinning of general relativity, which has gravity and inertial acceleration as indistinguishable locally. In the grand scheme of things, the cosmological theories are subordinate to quantum physics or relativity. Even advocates of the MOND theory are reluctant to pervert general relativity in support of it.

Since MOND has been proposed several alternative theories have been put forward that try to make it more compatible with general relativity, and explain galaxy configurations that stymie MOND. One such theory is TeVeS (tensor-vector-scalar gravity), a relativistic extension of MOND, proposed by Jacob Bekenstein. Another goes by the acronym MOG, developed by John Moffat, that is not only relativistic, but claims gravity varies, and not inertia. It also explains many more cluster galaxy arrangements. The important point is that all these theories dispense with dark matter and challenge the equivalence principle.

Another major bit of corroborative evidence, for the dark matter hypothesis is gravitational lensing, the bending of light by massive celestial bodies, which was envisaged by Einstein. The first confirming test of general relativity was the bending of light, manifested by the apparent shift of star positions, as starlight grazed the sun. A more recent test of gravitational lensing is the bending of light around clusters of galaxies. The basic idea is that the massive cluster can deflect light so that what an observer might view as the position of a bright object near it, typically quasars, may not be its real position. An intervening galaxy acts as a lens to focus the image of the distant quasar to a new location. Depending on the geometry of the cluster and the position of the quasar, multiple images or even a ring of images for a single bright star may be seen.

Earlier it was shown that light bends around the massive bodies, like the Sun, twice as much as expected by a simple application of Newton's gravitation principle. This occurs because of the duality of light's particle and wave-like properties. For odd shape bodies, the bending factor might be different for the two. In addition, the bending due to its particle-like qualities is maximized closest to the center of mass, but the bending due to its wave-like qualities is greatest on either side of the center of mass. These phenomena are separate from MOND, but obviously needs to be considered for the lensing calculation. For example, it might explain why the bullet cluster, two colliding galaxies, exhibits it's maximum lensing away from where expected with only baryonic matter and general relativistic thinking. In addition, unlike TeVeS, it would not have a stability problem, needing to reintroduce some dark matter into the mix.

The following figure illustrates lensing.

Figure 7.5 - Gravitational Lensing by a Galaxy

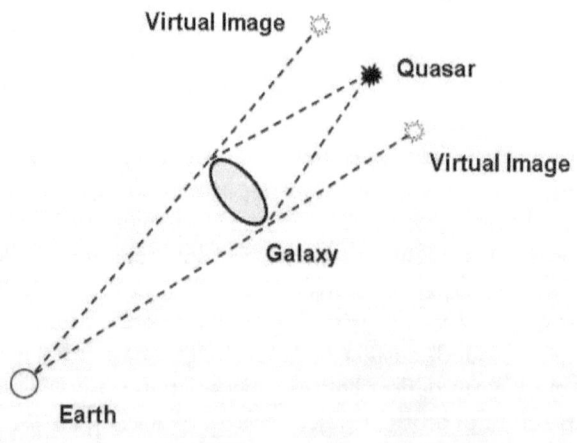

Figure_7.5 - This is a diagram to show the effect of gravitational lensing by a galaxy. The actual position of a quasar, one of the brightest stars in the sky, is situated behind the galaxy, perhaps 50,000 light-years across. Two virtual images of the quasar's apparent position are seen from Earth due to the bending of the quasar light around the galaxy. The identical red shift and spectra of the two images and their relative closeness to each other make it highly likely that the same hidden quasar causes both images. Knowing the angle between the two virtual images and the distance to the galaxy, the mass of the galaxy can be estimated. Subtracting this mass from the visible mass, the amount of dark matter in the galaxy is inferred.

If the distance to the cluster is known and the bending of light is measured, the mass of the cluster can be calculated. However, general relativity teaches us that this bending of light is not caused by a physical attraction of light to matter, but by light traveling through space distorted by massive objects. Using information about the rings and multiple images allows astronomers to reconstruct the total mass distribution. Subtracting away the luminous matter from the total matter provides knowledge about the distribution of the presumed unseen dark matter.

General relativity and dark matter are closely bound. Dark matter explains the galaxy's containment, even though the outer stars are moving fast. It also explains the degree of lensing, unexplainable with just the visible galactic matter. Curiously, if the dark matter hypothesis were wrong, it would inflict a serious blow to general relativity, since general relativity presupposes that the force of gravity and the force of inertia are intimately related to each other.

On the other hand, the gyroverse is based on a mechanical model of the universe where the equations are not sacrosanct, as with general relativity. For physical phenomena, transitions, including their derivatives, are not abrupt. The force generated by acceleration is a vector $F=ma$. Its derivative with respect to acceleration is the mass, m. It is also a directionally dependent vector having multiple vector values at each point in space, making it discontinuous. Thus, it is reasonable to expect that for a small interval this first derivative should also be continuous. $F=ma^2/a_0$, as proposed by MOND, provides such a first order approximation of the transition, since its first derivative with respect to acceleration is zero at zero acceleration, independent of the direction of the acceleration. More importantly, it only takes a small force to rein in the outlying stars. That is why MOND can explain the higher-than-expected velocity of stars far from the galactic center. In addition, it also explains the lensing of light around distant galaxies. Just as stars at the outer limits of the galaxy are accelerating very slowly toward the galactic center, the light rays grazing the galaxies also accelerate very slowly in its transverse inward direction toward the galactic center. As a result, both phenomena produce commensurate, but smaller amounts of inertia than suggested by general relativity. This allows for much more acceleration than the galaxy's gravitational force would suggest, compensating for the fact that the galaxies have only one tenth the quantity of matter otherwise needed.

An interesting byproduct of MOND is that it violates the conservation of energy principle, the cornerstone of all physics. At a very low acceleration, the speed of a mass could be increased using much less force than $F=ma$ would otherwise suggest. If the mass subsequently decelerated at an ordinary rate, the reclaimed energy would be greater than the kinetic energy originally imparted to the matter. In principle, this would allow for the creation of a perpetual energy machine. However, using this technique may be impractical, because it would take an inordinate amount of time to build up to an adequate speed. In addition, friction and wind resistance would eat away at any created energy. Recall that this is the second example of getting more energy out than what was put in, the "big push" being the other.

The upshot is that while MOND is counter-strategic to general relativity, it is especially consistent with the gyroverse theory. In fact, a smooth transition is expected of all mechanisms, including the gyroverse. Usually, physical models at their extremes break down tending to produce alternate analytical expression. Several other examples of this have been cited in this book. Gravity, which attracts locally, repels at the universe's outer limits. The electromagnetic force becomes only repulsive at subatomic distances. "Black holes" are not singularities. They produce a very large attractive force, nevertheless, one that is finite.

PIONEER MYSTERY AND ALLAIS EFFECT

Recently it has been observed that the *Pioneer 10* and *11* space probes, launched by NASA in the early 1970s, are receding from the sun slightly more slowly than expected. While the deceleration caused by the sun is much greater than experienced with the previous galactic phenomena, the magnitude of the acceleration error is roughly in the range consistent with the MOND theory. Since these probes are traveling in opposite directions away from the Sun, most other explanations had been discounted, until recently when asymmetrical thermal radiation has been revisited. However, the rapidity of acceptance by the physics community makes it suspect. The deviation in acceleration is about $(8.74 \pm 1.33) \times 10^{-10}$ m/s² for both spacecraft. Nevertheless, while other reasonable explanations are lacking, MOND is discounted by the mainstream because it is in conflict with general relativity.

Another phenomenon that mistakenly appears to be similarly caused goes by the name Allais effect. In 1954 Maurice Allais, a 1988 French Nobel Laureate in economics, observed that a paraconical pendulum experienced unexpected anomalies during a solar eclipse, when the moon passed in front of the Sun, indicating a drop in gravitational force. Whereas the change in pendulum motion is small, repeated testing, some with alternate instruments, seems to confirm the results. The conventional explanation, dubbed Allais effect, proposes that a large mass like the moon could block some of the gravitational force of the Sun behind it.

Clearly, this explanation would also conflict with general relativity. Consequently, there have been efforts to tweak general relativity slightly to explain both these perceived flaws in it. It is somewhat ironic that testing during a Sun's eclipse was instrumental in confirming general relativity, and now might be influential in disproving it.

While the Allais idea appears correct, it is the wrong metaphor for explaining the pendulum speed-up. Gravity is caused by the LeSage mechanism. When the moon eclipses the sun, it blocks fewer graveons from striking earth below, weakening the gravitational attraction on matter. Instead of causing a change in gravitational pull by blocking graveons, it causes it by blocking fewer graveons.

Figure 7.6 – Allais Effect

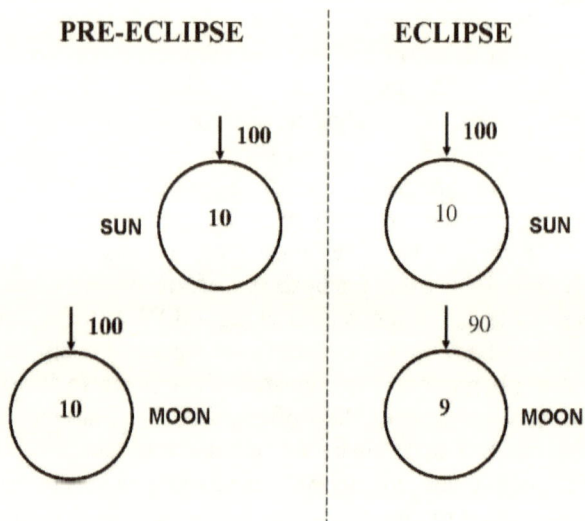

Figure_7.6 – The illustration on the left depicts the sun and moon just prior to an eclipse. As they move closer together, more of the combined vertical component of graveons is blocked increasing the gravitational "pull" on earth objects below. For illustration, purposes 100 units of graveons are assumed to impinge downward on each, absorbing 10 units of it that would otherwise continue down to earth. As gravity suddenly changes, a sensitive pendulum on earth below reacts to the change.

On the right-side illustration, the moon eclipses the sun. The moon no longer receives the 100 graveon units, but instead only 90, blocking 9. Since the sun and moon together only block 19 graveon units, the force of gravity decreases during the actual eclipse and the pendulum speeds-up accordingly.

In order to understand this, consider the following illustration. Just to make computation simple, assume the Sun and Moon each block 10% of the graveons that hit them. If 100 graveon units strike each one, each would block 10 units and pass the rest. If the Sun and Moon were adjacent, jointly they would block 20 graveon units from striking Earth. If the Moon eclipses the Sun, the Sun still would block 10 units, but 90 units would pass through and strike the Moon. Since the Moon would block 10% of the graveons that strike it, 9 graveon units would be blocked by the Moon, making 19 units the total number blocked. This is one less than when they were adjacent and would reduce the gravitational attraction accordingly.

FLYBY ANOMOLY

The flyby anomaly is the unexpected small change in energy of satellites passing very close to earth. This type of maneuver is used to give a gravitational assist to spaceshots destined for other planets in our solar system. While the speed relative to earth is not materially change, the speed relative to the destination planet does, saving time, and using much less fuel. This technique was discovered by Michael Minovich, a graduate math student, working summers at the Jet Propulsion Lab. Mariner 10 in 1973, headed for Venus and Mercury, was first to use this procedure.

Contrary to expectations, when the perigee of the flyby trajectory is near the Earth's equator, the spacecraft's speed when exiting the flyby is slightly different, albeit a small amount, in the neighborhood of one part in a million. If the entering path is closest to the equator, then the exiting speed is greater than the entering speed, but if the exiting path is closest, then the opposite happens.

Current physics is at a loss at explaining this phenomenon. Atmosphere drag, general relativity effects, and dark matter have been ruled out. In the Gyroverse theory, space as we experience it is four-dimensional. When an object moves, its position space occupies a different three-dimensional subset of space. Since earth at the equator rotates at about one thousand miles per hour, its three-dimensional position space subset by the equator is different by about one part in 625 thousand. Thus, gravity is 'pulling' to a slightly different point than would be from a static Earth in the same position, causing a small difference in the speed and trajectory. If a spacecraft's pericentre is close to the equator, the speed difference is minimized, but the trajectory difference is maximized.

BONDING ENERGY

Current theory posits that a gravitational field contains negative energy because, two masses that are close together have less energy than if far apart since it takes energy to pull them further apart. One theory goes one step further and speculates that the universe has zero net energy. Its negative component is the energy of gravity, and its positive components, which completely offset the negative, are the energies of motion and matter.

While the argument that gravitational field is negative energy is computationally accurate, it is incorrect because energy is never negative. It is just that the sum of the masses of two objects is less if the objects are closer together, making mc^2 smaller, while leaving the amount of matter the same. This same principle applies to atoms since it too is held together with graveons. The reason for this follows easily once it is remembered that the mass of an object is proportional to the amount of graveons captured by the object.

Figure 7.7 – Bonding Mass Decrease

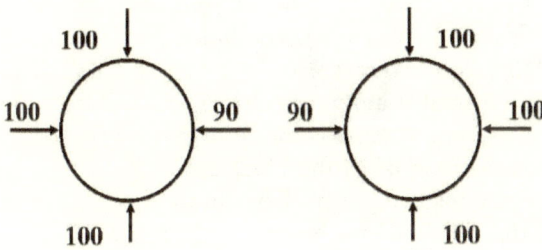

Figure_7.7 – Both of these objects are nominally bombarded in each direction by 100 units of graveons. Including 100 units in and out of the paper, the total showering each would be 600 graveon units, if not for the fact that each object blocks some graveons from impinging on its neighbor. Not only does the imbalance of graveons drive the objects together but the total hitting each is also reduced. Since the mass of an object is proportional to the number of graveons captured by it, the total mass of combined decreases as the objects are brought together.

It must be remembered that the graveons are actually traveling in twelve-dimensional space. but nothing is lost showing it this way.

To illustrate this assume there are two objects standing alone far from other objects that are showered with 600 graveon units (including 200 units in and out of the paper). If these two objects were brought close enough together, each one would block 10 units from striking the other object. Under this scenario the number of units striking each of these objects would be reduced by 10, bringing the number of hits 590 to each. Since the objects capture 10% of the graveons striking it, each will only capture 59 graveon units. The result is that if the two objects are far apart, together they capture 120 units, but when brought close enough together only capture 118 units, reducing its mass accordingly.

Since large objects and small atoms are held together similarly with the exact same mechanism, this description applies to both.

For example, protons and neutrons separately have more collective mass than larger atoms with several nucleons. The Hydrogen bomb uses this principle, converting Hydrogen with one nucleon to Helium with four nucleons, reducing the mass and consequently its mc^2 energy. The mass of the two protons and two neutrons separately is greater than when combined into helium. The excess energy of the conversion creates the huge blast. The process of combining the smaller atoms into larger ones is termed fusion, the opposite of fission. When it comes-to much larger atoms, elements high on the periodic table, the blocking of graveons is not only a function of the many other nucleons in the atom but is dominated by the configuration of the nucleons. If they are bunched tightly together, they block more graveons from striking each other, than if packaged more non-uniformly. This is the principle of the fission bombs that uses Uranium. The fission products are packaged denser than the original Uranium atom, blocking more graveons from striking each other, making up for the fact that the Uranium atoms are being broken into separate parts. This type of process usually occurs through beta or alpha decay. The process is stopped when the resulting isotopes have less collective mass than neighboring atoms on a table of isotopes. This explanation is slightly modified by the electromagnetic repulsive force that works to break the molecule apart. However, since photons also get their inertial power from graveons, the net result is the same, even though that exact mechanism is not obvious.

STELLAR ABERRATION

Stellar aberration, first discovered by James Bradley in 1728, is the apparent shift in position of stars due to the motion of Earth as it orbits the Sun. Contrary to the basic tenet of special relativity that absolute motion does not exist; stellar aberration is not affected by the motion of stars, but only of Earth. Even stars with orders of magnitude more lateral motion than Earth's orbital motion, cause no aberration. Ironically, special relativity, with its flavor of Lorentz transformation, originally used aberration to prove the efficacy of relativity. Then, it was believed that the universe was static, the stars fixed in place. Apart from that, relativity explained the apparent constancy of the speed of light for radial motion and aberration for lateral motion better than all variations of the aether hypothesis. Had it been known then that binary stars has immense relative lateral motion in relation to Earth, the discussions might have taken on a completely different twist.

Bradley found the aberration to be about 20.5 arc-seconds, which was important evidence supporting Copernicus's heliocentric theory that Earth was orbiting the sun. Using that measurement, he was further able to calculate the speed of light. Bradley was born in England, became a professor of astronomy at Oxford, his alma mater, after a brief career in the clergy. In 1842, he was appointed Astronomer Royal at Greenwich Observatory, modernized it, and continued with his stellar observations. After a long surveillance, in 1748 he announced the discovery of nutation, a wobbling of Earth's axis that has an 18.6-year period.

Stellar aberration is a very important phenomenon, because it is in clear violation of special relativity. The heart of special relativity's mantra is the postulate of no preferred inertial frame; absolute motion does not exist. Accordingly, it should not reveal whether the light source or the receiver is doing the moving--apparently, sometimes it does. It is amazing how many relativity authorities are mute on this point. For the gyroverse theory, no such restriction exists.

The gyroverse theory, with its own variation of Lorentz transformation, also explains the constancy of the speed of light and aberration, without abolishing absolute motion. It was already discussed that celestial light approaching Earth, gradually takes-on the prevailing motion-direction as it spirals toward us, hugging the 3-sphere. Aside from Doppler shift, light from binary stars, for example, reverts to the same motion direction, not exhibiting other evidence of relative motion. This hides stellar lateral motion. The apparent position of stars is nothing like where the light originated.

The orbital speed of Earth is about 30 thousand meters per second. This reflects a position-space shift in gyrohelix yaw for Earth's coordinates compared with a hypothetical stationary Earth of 20.5 arc-seconds. Galilean and Lorentz transformation are not substantially different in the lateral direction. For example, a 10-thousand-meter change in position-space will at most result in a meter difference between both transformations. This small difference will not affect lateral directional measurements between the two appreciably; so, whatever aberration is calculated from Galilean considerations will also apply to the gyroverse.

DELAYED CHOICE EXPERIMENT

Neils Bohr said, "Anyone that is not shocked by quantum theory does not understand it." John Archibald Wheeler, colleague of Neils Bohr and Albert Einstein, extended this absurdity one step further by proposing the Delayed Choice Experiment. Wheeler, previously coined the term black hole, and was a major player in the development of the atomic and hydrogen bombs. One of his major contributions was his proposed variation of the double-slit experiment in which the configuration is altered, on the fly, after the photons would have already passed through both slits. In the first variation, an interference pattern results, but in the other configuration, no interference pattern results and it is possible to identify which of the two open slits the photon actually passed through. In effect, applying orthodox quantum theory, the experimenter seems able to alter the past. Furthermore, Wheeler proposed a cosmic scale version of this experiment where the past alteration does not just go back a fraction of a nanosecond, but millions of light-years.

The following figure is an implementation of John Wheeler's delayed choice experiment on a cosmic scale. In this thought-experiment absurdity, light from the quasar straddles the intervening galaxy. These two distinct paths are equivalent to the double-slit configuration, shown earlier. The telescopes point to the quasar light to insure no stray light from surrounding areas interferes with the experiment. It is assumed that the quasar is far enough away so that light arrives one photon at a time. On the left drawing (A), photons are assumed to arrive at the silvered edge (shown as a black line) of the half-silvered mirror from both directions simultaneously. The half-silvered mirror reflects half the light that impinges on it from either direction and passes the other half through it. In effect, the reflected light bounces off the silvered side, shown as a black line. When light does not have to pass through the glass to impinge on the silvered side, the reflected light experiences a phase reversal (half wavelength shift), without a corresponding time delay. When it must pass through the glass to reflect from the silvered side, the phase reversal is absent. Adding up all the phase shifts, the reflected light on the far-left detector has a half wavelength more phase shift than the light transmitted through the glass to the detector. As a result, no light hits this detector, since the phase shift through the glass for the reflected light and the transmitted light is 180 degrees out of phase, destructively interfering with each other. For the detector on the right, both the transmitted and reflected light experience no phase reversal and the light constructively interferes, unabatedly passing photons to the detector.

The drawing on the right (B) is identical to (A) with the half-silvered mirror removed. No destructive interference occurs, and the firing of a detector shows which path the photon took. In this case, half the photons seem to approach from each side of the galaxy. Applying orthodox quantum theory, the left side (A) configuration signifies that the photons had to travel on both sides of the galaxy to interfere with itself destructively. On the other hand, the right side (B) configuration indicated that the photons only traveled on one side or the other of the galaxy. Since it took millions of light-years for the photons to have reached Earth, the theory implies that the experimental configuration had to alter the past in order to have allowed for both outcomes.

In the gyroverse explanation, all the photons are always traveling in sister paths, which can be far away in three-dimensional space, but exceedingly close in twelve-dimensional space. By altering the configuration, which detector the photon finally strikes is selected by chance at the last moment, without the need to change the past. Unlike orthodox theory, as sister-locations are separated further apart in three-dimensional space, this phenomenon starts to fail. It is not clear if it would work for this cosmic size configuration.

As an aside, this is interesting, because the fine structure constant is based on the assumption that every photon path is as likely. Only the fact that the more roundabout paths of the Feynman Diagram combine destructively, do these extreme paths not contribute much to the result. It would be instructive to reconcile the difference between the calculated and measured value of the fine-structure constant by allowing distant paths to be slightly less likely to occur. Perhaps, this difference could also be used to estimate the distance limits of entanglement.

Figure 7.8 – Cosmic Scale Delayed Choice Experiment

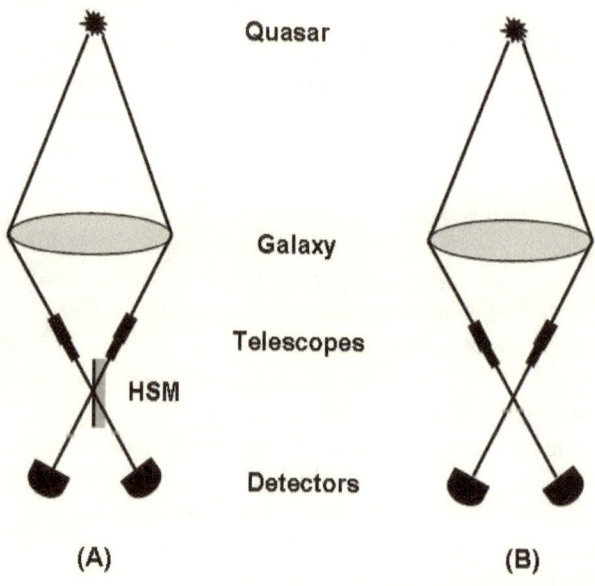

Figure_7.8 - This depicts double-slit thought-experiment of cosmic scale proposed by Wheeler. The quasar is the light source that is far enough away that photons are individually being detected. Light bending around a galaxy at both extremes represents the double-slits. Telescopes ensure that the focus is confined to the light from the quasar. Figure (A) allows light from both directions to interfere. On the left side of (A), the light reflected from the mirror has a phase reversal as compared to the light transmitted through the mirror from the other side of (A). This causes the light on the left detector of (A) to interfere destructively, resulting in no photons reaching it. On the right side of (A), the reflected light does not have a phase reversal, so that constructive interference between it and the transmitted light develops, causing many photons to be detected. This imbalance of detected photons illustrates that the photons on the (A) side are interfering with themselves. Notice that the light emanating from the right side of the galaxy always passes through an additional thickness of glass. This can be compensated for by having the light from the left pass through an equivalent thickness of glass before reaching the half-silvered mirror. On the (B) side, with the half-silvered mirror removed, both detectors get the same number of hits and according to orthodox theory it is possible to identify which side of the galaxy the photon came from.

While this was an interesting thought-experiment, it would not work. For one, quasars and galaxies are so distant that the two paths are surely, months, or longer apart from each other. However, laboratory versions of the delayed choice experiment have been successfully carried out in the mid-80's at the University of Maryland and the University of Munich. These implementations were variations of Wheeler's 1978 proposed delayed-choice double-slit experiment. According to orthodox theory, the experimenter is participating in creating the past. Again, in the gyroverse model, changing of the past is not needed to explain this seemingly bizarre result.

The following figure using a modified Mach-Zehnder Interferometer depicts the actual implementation chosen to demonstrate the delayed choice effect. This is the same Mach known for the Mach Principle that provided inspiration for general relativity's equivalence principle, and Mach number, the ratio of the speed of aircraft to the speed of sound.

Light emitted from the photon source splits equally along path 1 and path 2. The lengths of air and glass along both paths are equal on entering each detector. Now, the reflected light into D1 from path 1 experiences one phase reversal making it 180 degrees out of phase with light from path 2 which experiences two phase reversals before entering into detector D1. Thus, detector D1 never fires because the interference is always destructive. On the other hand, along path 2 there are two phase reversals into D2, matching exactly the two phase reversals from path 1 through H2 into D2. Accordingly, constructive interference at D2 causes it always to fire.

Donald Wortzman

This implies that that each photon must be traversing both path 1 and path 2. Otherwise, each detector would have fired half the time.

Figure 7.9 – Delayed Choice Experiment using the Mach-Zehnder Interferometer

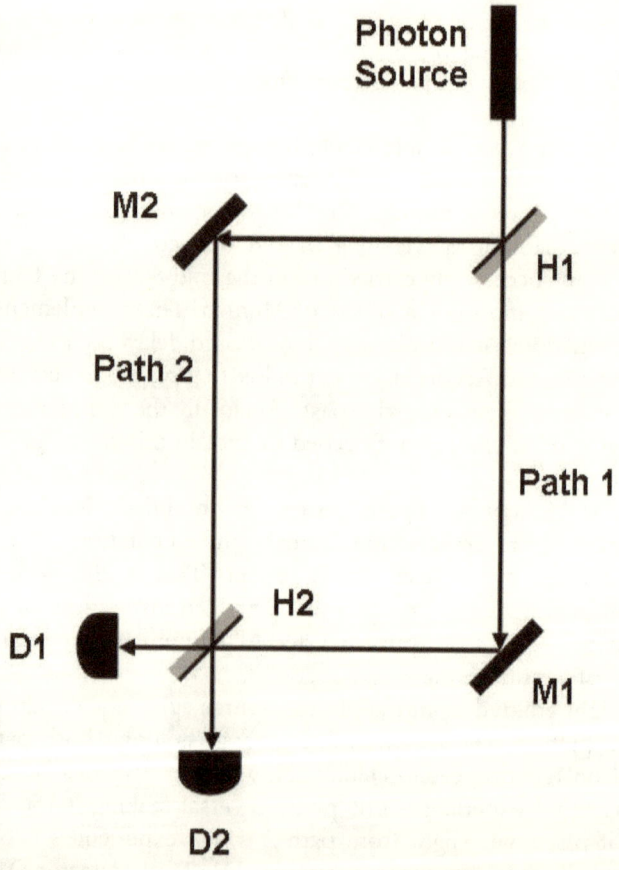

Figure_7.9 - This laboratory version of the previous cosmic scale thought-experiment has been successfully carried out. As shown, this figure is equivalent to Figure_7.7 (A). D1 never fires and D2 always fires, giving the appearance that each photon is interfering with itself. With the half-silvered mirror, H2, removed, the configuration defaults to the equivalent of the (B) side, where both detectors fire half the time appearing as if one could distinguish which path the photon has taken. If removing or replacing the half-silvered mirror, H2, occurs before the photon would have reached H2, the switch gives the appearance that the photon has gone back in time to cause the change.

To demonstrate the delayed-choice aspect of this setup, the half-silvered mirror H2 is removed after the photons are calculated to be beyond M1 and M2, but before they impinge on H2. When this is done, detectors D1 and D2 each fire exclusively half the time, showing that the photons took only one of the paths. This changeover occurred after the photons passed M1 and M2, signifying, according to orthodox theory, that the photons had to go back in time to make this change.

Again, the gyroverse's explanation finesses this apparent contradiction to explain this seemingly bizarre result. Photons as well as all other particles can be distributed over vast distances while really being exceedingly close in sister-locations. Where a photon shows up in three-dimensional space is not decided until the very last moments, by chance. In effect, it is always traveling down all paths simultaneously, with just a different physical probability of showing up at each location.

DIMENSIONS OF MATTER

It is interesting to ponder whether three dimensions are special. Could a universe be defined where all objects have a different number of dimensions, perhaps two or four? Most of the known physics parameters are limited to three dimensions. If something is two-dimensional, it has no thickness. If it has no thickness, it has no mass, and for all practical purposes is not perceivable except in thought. A shadow comes closest to a two-dimensional object, but it needs a three-dimensional reflective surface to be seen. On the other hand, just as a four-dimensional hyper-box has infinite volume, a four-dimensional object would also have infinite volume, and thus, infinite mass. Clearly, this too is not possible. Every object of matter in our universe is exactly three-dimensional--no more, no less. Likewise, inertia, gravity, and energy as we know them are also strictly three-dimensional phenomena, for in more dimensions they would be infinite. The measure of time regulated by inertia is also three-dimensional. In a different number of dimensions, none of the laws of nature relevant to that number of dimensions would be recognizable to us.

HIDDEN VARIABLES

One shortcoming of physics is the over confidence placed on proofs. Proofs are only as good as the assumptions, both explicit and implicit upon which they are based. Einstein called entanglement "spooky action at a distance," and believed that quantum theory was incomplete with hidden variables that would eventually explain the phenomena. Neils Bohr and Einstein had a longstanding disagreement on entangled particles, discussed in more detail in other parts of this treatise. Quantum theory, which Neils Bohr strongly endorsed, predicted that entangled particles would remain entangled even if huge distances separated the particles, remaining cognizant and reflective of each other. Einstein believed it only appeared that way because we did not have a complete understanding of the physics that he dubbed hidden variables. After both died, Bells inequality and Alain Aspect's experiment (discussed later) convinced the physics community that Neils Bohr was correct, and Einstein was wrong about the existence of hidden variable--that is just the way nature works.

The scientific community took too narrow a view of the definition of hidden variable. Hidden variables include the hidden physical structure of the universe, which enables particles that appear distant to each other in three-dimensional space to be exceedingly close in the full space. The assumption that two objects seeming far apart are actually far apart is so obvious that it does not need mentioning--except this theory shows that the assumption can be wrong.

It is reminiscent of the worker on a construction site that left each day with a wheel barrel full of sand. The guard searches through the sand each time, but never found hidden tools or any other contraband because the employee was stealing wheel barrels. If the employee had left with empty wheel barrels, he would be caught immediately, but adding sand redirected the guards thinking away from the wheel barrel.

The phrases hidden variable, or incompleteness of quantum theory, implies bounds on the nature of the missing information. The missing information was a hidden assumption about the nature of the hidden variables. Even if this gyroverse theory ultimately is wrong, the fact that a model can be envisioned where entanglement would not be strange opens the possibility that other models, as well, could explain this and all the other strangeness in the universe.

CMB Dipole Anisotropy

The first satellite to accurately measure the background radiation, remnant of the formation of atoms epoch, where light could travel unencumbered was the COBE, Cosmic Background Explorer satellite, launched in 1989. In 2001 another satellite, WMAP, the Wilkinson Microwave Anisotropy Probe was dispatched, which more accurately determined that the spectrum is consistent with a blackbody of temperature 2.725 +/- 0.002 Kelvin. More important it measured the anisotropy, the maximum difference in apparent temperature from radiation in opposite directions. This small Doppler shift, three orders of magnitude smaller than the nominal frequency, is a measure of the motion of the satellite relative to an isotropic cosmic microwave background, CMB. After subtracting out anisotropic dipole component, what remains is the expansion velocity of the 3-sphere, which is the minimum gamma inertial frame. The gamma on earth is somewhat larger, and the gamma of the WMAP satellite is still larger. The satellite orbiting Earth, Earth orbiting the sun, the rotation of the Milky Way, and the local cluster sailing through space contributes to this Doppler shift. Nevertheless, by some estimates, all of these contributing factors are not enough to explain the anisotropy.

Just after the big push, all expansion was radial. As the uniformly distributed matter congealed into celestial object, Newtonian dynamics insured that motion remained essentially radial. However, MOND, modified Newtonian dynamic, makes no such promise. Consequently, it is highly likely that the local super cluster has a net motion slightly non-radial. To illustrate this point consider the following example. Assume two celestial objects, one twice as massive as the other, are attracted to each other. If each one is moving radial to the 3-sphere, their joint center of gravity is also moving radially outwards, at the speed of light. Gravity causes the smaller one to move twice as fast as the heavier one toward their center of gravity, giving them equal and opposite momentums. The center of gravity stays fixed, even if the objects crash. If the expansion of the 3-sphere starts out as radial motion, Newtonian dynamics encourages it to stay so. However, for MOND dynamics, the situation is different. More massive objects, which generally accelerate more slowly, will move faster than Newtonian dynamics demands. This will cause the center of gravity to move toward the smaller object, giving it a non-radial component. This in effect will give local celestial matter, up to the size of super-clusters, a net non-radial component. Since the radiation was emitted when all matter was still moving purely radial, the non-radial motion today causes the background radiation to exhibit some anisotropy.

OTHER UNIVERSE CHARACTERISTICS

From outside the universe, matter has little or no inertia, mass, or energy, since these phenomena are internally generated from the internal gravitational forces. Even the graveons that generate the gravitational force may be much less substantial than presumed. Since the graveons are pushing against matter that has few intrinsic qualities, it does not take much force to do the job. That is why the universe is less formidable from the outside than from the inside, especially at its beginning.

Theories hypothesize that the total energy of the universe is zero. The reasoning is that the total energy of the universe is counterbalanced with the gravitational field energy, which surprisingly is negative. Guth's Inflation Theory uses this principle to hypothesize that the universe began with little matter and that this matter energy spontaneously materialized, counterbalanced by negative gravitational energy. It is intriguing that it can be shown that gravitational energy is negative, but it cannot be proven using the traditional universe model that it is sufficiently negative to counterbalance the universe's positive mass energy.

Another ponderable is time, which is paced by inertia. The more inertia, the more sluggish everything becomes from atomic reactions to mechanical motion. Without inertia, these reactions would progress more rapidly, in fact instantaneously. Everything would happen all at once, perhaps, immediately being in its final state. Inertia throttles time like a spigot, letting it out bit by bit. While this slowing of time is comprehensible, a place without time is not easily imaginable. Without time, events cannot have an order, an end, or a beginning. Exactly what the notion of time would be like outside the gyroverse model is mystifying. The time-like measures outside are surely altogether different from what they are on the inside.

CONCLUSION

While the universe is vast from our point of view, in a twelve-dimensional multiverse, it is the equivalent of one atom and quite small. It is likely that many more universes reside in this space. A twelve-dimensional box, an eighth of an inch on each edge could contain more than 10^{80} of these universes--more universes than atoms in this entire universe. Especially fascinating is that the universe has an outside and is finite. From the inside, it is difficult to imagine it either being finite or being infinite.

Would intelligence on the outside even be aware of the gyroverse? The gyroverse has atomic size dimensions. An intelligent entity might be comparatively larger than the gyroverse, as we are to the atom. It may not be aware of the gyroverse. After all, we are in the dark about the inner conditions within atoms. None of our senses is operational at the atomic level. The things we do know about the innards of atoms are at very course granularity. Similarly, the universe might be evolving without their cognizance.

As this theory evolved, it became increasingly evident that it was more probably the product of careful design. The design appears too clever to be accidental. Its twelve-dimensional construction enables a tightly wrapped bundle of matter to appear infinitely large. Even though celestial bodies were much closer together in the past, and rushing away from each other at the speed of light, the spiral path of the primeval celestial light makes the universe appear uniform, Euclidean, and static. Just about all the visible matter in the universe can be seen, with perhaps some of it, seen twice. The forces of nature have not serendipitously assumed proper proportions for intelligent life, but they were derived from a single mechanism, caused by the graveon. The inertia generation and amplification of the gyroverse mechanism keeps everything in the universe stable with respect to everything else. It constrains all but the tiny subatomic particles only to take the long circuitous path everywhere. While the universe is very massive from the inside, it is really a tiny fraction as substantial from the outside. The energy of matter is not intrinsic to matter but is caused by the gyrohelix working against the graveons. The gamma factor keeps the look and feel of the universe constant compensating for the absolute changes in the physics parameters as the universe evolves. These things, and more, add credence to the postulate of an intelligent design. It is amazing how this much sophistication is achieved with so few elements. Starting with graveons, photons, and matter, which are all wave packets moving in tiny circular orbits in a twelve-dimensional space, all the characteristics of the universe have been created. These include all the forces of nature, all the mass and energy in the universe, an almost infinite Euclidean space, and all the other complexities. It could be an accident of nature, but it is doubtful, because it appears to be so cleverly designed.

The gyroverse's physical structure of the universe is a wide departure from the traditional universe model, having a very significant impact on all aspects of modern physics. To present a complete picture, some speculative aspects were included in the treatise. It should be judged on its entirety, and not on each individual item. The overwhelming evidence for the model is the many anomalies that cannot be explained with traditional theory.

Lastly, as advised in Chapter 1, an old bromide for making something understandable says: "First, tell them what you will tell them. Then tell them. Finally, tell them what you told them." In keeping with that tradition, I submit the following.

STRUCTURAL EVOLUTION OF THE UNIVERSE

A little over a hundred years ago, physics took a very unfortunate turn. It went from being a physical science, as the name implies, where it was couched in terms of how nature worked, to being mostly mathematical, with almost no regard for how nature behaves. A new theory, the gyroverse, takes physics back to the physical, thereby modifying many aspects of current accepted physics, explaining nature better.

I will be discussing here one of the most important aspects of this theory, Structure and Evolution of the Universe. Understanding it will go a long way in understanding everything else about the theory. Starting from the big bang, I will progress through the expansion of the universe to the present time.

In short, the universe is on the surface of a four-dimensional ball expanding at the speed of light. To aid the discussion I am using the following illustration. The illustration depicts the universe from the big bang to the present. While it's drawn in two dimensions, it represents the four-space dimensional view of the universe. The revelation that the universe has four space dimensions is new to this theory. Each of the concentric circles represents the universe at billion light-year intervals. The circle labeled 14 depicts the universe today, 14 billion light-years after the big bang. The one labeled 1 represents the universe, as it was one billion light-years after the big bang...and so on. As the universe expands galaxies get further from each other in proportion to their distance apart, making it appear that empty space is being created, as assumed in current theory. However, celestial object close to each other become gravitationally bound together and eventual expand as one in the same direction.

Again, the entire universe is on the surface of the expanding four-dimensional ball, called a 3-sphere. The expansion direction represents one space dimension, and each circle, one billion light years apart, represents the three-dimensional universe as it was in its respective epoch. The expansion velocity "c" gives all matter the kinetic energy, mc^2. Matter just moving in the radial direction is at its minimum kinetic energy level, or at its rest state. Hence, matter, itself, is not intrinsically energy in another form. If the expansion ended there would be no energy in the universe, since the expansion energy is what fuels all forms of energy in the universe. For example, the changed total mass, in any chemical or nuclear reaction, moving at the speed of light on the surface of the universe, is the source of the energy for that reaction. All energy is kinetic.

EVIDENCE FOR MORE THAN THREE DIMENSIONS

There is some evidence that suggests the universe is made up of more space dimensions than three. Recently, the furthest galaxy GN-Z11 was observed, as it existed 13.4 billion years ago. It has Z factor of 11.1, which equates to a recession speed of .986c, making it 13.2 billion light years away as observed. Looking further in that direction, where the universe was opaque, the CMB has a Z factor of greater than 1000, implying a recession speed of .999998c.

It is too much of a coincidence for the CMB's Z be greater than 1000, since there is no a priori reason for the expansion of space between the CMB origin and the Milky way to be so close to c. It would be even a bigger coincidence, if other galaxy's recession speed with their CMB were not also very close to c, because that would make earth a very special place.

If all galaxies are equidistant from the CMB origin, it would imply that all galaxies are on the surface of a 2 dimensional sphere. That's clearly not possible since the galaxies are uniformly distributed in 3 dim space, as visually seen. Consequently, the universe need be more than just three space dimensions.

If one adds the fact that CMB is the same in every direction, a three-dimensional universe is even more problematic

Figure 7.10 – A Time Slice of the Universe

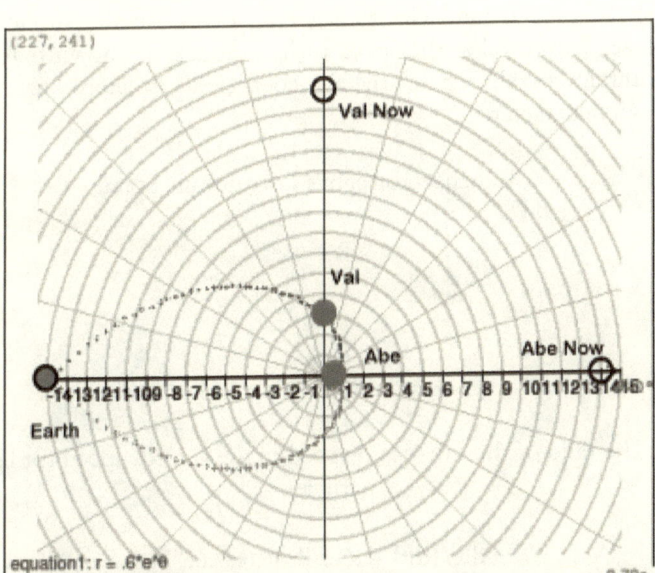

Figure_7.10 - This is a time slice of the universe. It shows the center point where the universe began as a sub-atomic 4-dimensional box of every particle currently present in the universe. Amazingly, it fit in with room to spare. The universe had expanded with each concentric circle representing a billion light-years. Now 14 billion years after the big bang, it shows the path of light emanating from two galaxies toward earth. It also shows those galaxies today.

This illustration can be viewed as a time slice of the universe. It depicts not just the universe today, but also, it's form way back to the big bang. The center of the concentric circles is where the big bang happened 14 billion years ago. Interestingly, the site of the big bang is made up of 4 space dimensions of sub-atomic proportions, which is very roomy even though it is tiny. Strange as it may seem, every atomic particle in the universe could have occupied that tiny original space with no two touching each other.

For now focus on the three objects depicted on the blue dotted line, lights path, a logarithmic spiral of the form $R = .6e^{\theta}$, in polar coordinance (R,θ). Knowing that the expansion of the universe is at the speed of light and electromagnetic waves, on the surface of universe, also travel at the speed of light, it can be shown that the logarithmic spiral, where theta's coefficient is unity, has the necessary characteristic.

Equation for light's Logarithmic Spiral

$$R = .6e^{\theta}$$

and

$$dR/d\theta = .6e^{\theta}$$

The ratio of the expansion speed to the surface speed for light particles

$$dR/Rd\theta = .6e^{\theta}/R = .6e^{\theta}/.6e^{\theta} = 1$$

This shows that the expansion speed divided by the surface speed of light is equal to one, thereby making them the same, demonstrating that this logarithmic spiral has the correct characteristic. The .6 was chosen, for convenience, so that Abe could be on the plus x-axis, and Earth be on the negative x-axis.

On the far left is Earth now, 14 billion light-years after the big bang. Just above the center of the chart is the galaxy Val, shown when the universe was only 3 billion light-years old. And to the right of the big bang location is the galaxy Abe, showing where it was about 6 tenths of a billion light-years after the big bang. The blue dotted line is the path of light coming from Abe and Val to Earth. The look back to Val is approximately 11 billion years ago, and to Abe it is approximately 13.4 billion years ago. The light from Abe reached Val about 11 billion years ago, and then together with light from Val continued to Earth.

Notice that the red dotted curve also connects Abe to Earth. In effect light from Abe reaches Earth along the red line as well as the blue line. In other words, it is possible to see Abe from opposite directions. In fact, any galaxy, old enough, on the other side of the universe can be viewed from opposite directions, maybe even from multiple directions. In fact, the CMB which is about the size of a large galaxy, and further back than all of them can be seen in every direction, receding almost at the speed of light. Abe and Val are also shown on today's sphere, where it is 14 billion years after the big bang.

Interesting, in principle, it should be possible to prove the theory using that fact that galaxies could be seen from multiple directions. But, an object that far away is very faint, and not easy to identify especially viewing it from different directions. In addition, the distortions caused by lensing, necessary to view such a distant object, would make it look very different from different directions. There would have to be something very unique about the galaxy or its cluster to be able to identify it from different directions. The path that light took was determined, on the one hand, as photons traveling on the surface of the universe shell at the speed of light, and on the other hand, as the surface of the shell is likewise expanding at the speed of light. When both items are added to the fact that the universe is presently 14 billion years old, the equation for the path of a light-beam can readily be derived. On the left bottom of the chart are the resulting logarithmic spiral equations for the red and blue light-beams.

Notice that if those equations were continued back in time, they clearly would circle the universe many times. So in the earlier years, all radiation circled the universe much more often. For example, when the universe was just one hour old, radiation from anyplace would have gotten everywhere else in less than 25 hours.

Or similarly, when the universe was one thousand years old, it would have gotten everywhere else in less than 25 thousand years. So, unlike current theory, it isn't strange that the cosmic microwave background from all over the universe is the same. There was ample time for the temperatures throughout the universe to stabilize, prior to the formation of atoms, some 400 thousand years after the big bang, when the cosmic microwave background originated. Gravity, for the most part acts orthogonal to the expansion speed. So that the center of mass does not change as matter congeals. If matter started out uniformly distributed, it maintained its uniformity as the universe grew. Consequently, inflation theory is not needed to explain the uniformity. Also, unlike relativity, the distribution of matter has no bearing on the flatness of the universe.

The speed of light on the surface of the ball, or universe, is the same as the expansion speed. Just as with special relativity, the speed of matter is also restricted to the speed of light. It is the universe's expansion speed, which is the underlying limiting factor. Why that is the case is not obvious and was discussed previously. A photon is actually made of matter, whose rest mass is very tiny, but dispatched with a relatively large amount of energy to have it bump up against the light-speed barrier. In other words photons don't quite travel at the speed of light. The lower frequencies travel slower, but the speed difference is not yet measurable. If it were, the photon's rest mass could be estimated.

Notice that the path that light takes is curved, hence longer than linear. That is the reason far-off galaxies appear fainter than expected, which is mistakenly suggestive of an accelerating expansion. Because gravity between objects happens orthogonal to the expansion direction, gravity did not slow the expansion and congealed matter much easier. This enabled the expansion to be more gradual and uniform, and less of a bang. All the particles tended to stay on the surface of the sphere for two reasons. First the mass of expansion is the same as the mass of inertia, so that the expansion speed for all mater was the same. In addition, if any celestial objects strayed, the gravity of all the others nearby would pull them back onto the surface. Consequently, the expansion is self-correcting.

An interesting speculative digression is the attempt to unify the forces of nature that was made by Theodor Kaluza in 1921, possibly building on an idea from Nordström. He showed that starting with a theory of general relativity in five space-time dimensions, a four-dimensional theory of Einstein's general relativity plus Maxwell's electromagnetic theory would result. Oskar Klein made an important modification to the five-dimensional theory in 1926. He postulated that the extra dimensions are physically real, and that they could lead to the observed forces beyond gravity when looked at from our four-dimensional world. For many years, Klein attempted, in different ways, to promote the fifth dimension as a convenient way to introduce the quantum of action into the gravitational field. Despite Einstein's early encouragement and recognition of Kaluza's good idea, it seemed that Einstein was not convinced of the efficacy of the five-dimensional hypothesis, because the extra space dimension seemed contrived, just to get this interesting result. However, because the gyroverse theory contains a well-defined extra space dimension in its Euclidean view, it would make sense to revisit the Nordström and Kaluza-Klein Ideas, especially since time in GR when multiplied by c is the analog of the gyroverse's expansion at the speed of light. This could be a stepping-stone to the development of the gyroverse theory. For example, moving objects appear shorter because they are partially hidden in the fourth space dimension, and not the time dimension.

Another interesting observation that can be gleaned from this illustration is that individual fermions were taken to have spin 1/2, but should be 1 because during the universe expansion, each revolution of the gyrohelix causes each fermion to make one complete 360 degree rotation. Photons also complete one rotation at the same time due the universe expansion. However, photons also travel at the speed of light on the universe's surface, thereby possibly doubling the number of complete turns as compared to fermions giving them spin 2. The graveon has spin 0, because it doesn't go around with the helix, but cuts across the helix.

Figure 7.11 – Inside vs Outside Slice of the Universe

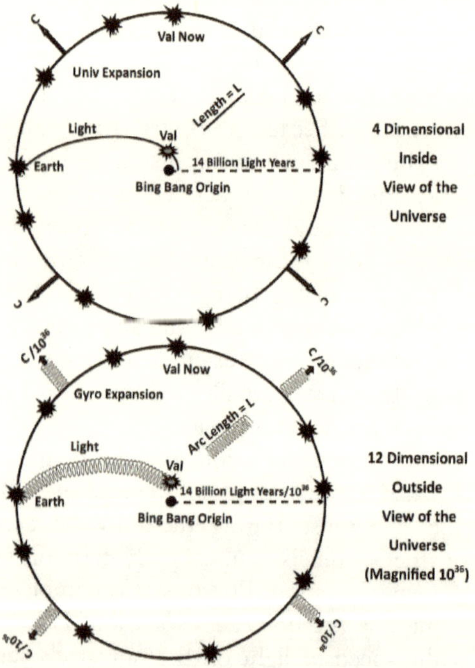

4 Dimensional Inside View of the Universe

(Labels: Val Now, Univ Expansion, Light, Val, Length = L, Earth, 14 Billion Light Years, Bing Bang Origin, c)

12 Dimensional Outside View of the Universe (Magnified 10^{36})

(Labels: Val Now, Gyro Expansion, Light, Val, Arc Length = L, Earth, 14 Billion Light Years/10^{36}, Bing Bang Origin, $c/10^{36}$)

Figure_7.11 - This shows two illustrations. The top one is a slightly different rendition of a time slice of the inside view of the universe. It also shows the center point where the universe began as a sub-atomic 4-dimensional box of every particle currently present in

The Gyroverse
The Hidden Structure of the Universe

the universe. The universe had expanded with each concentric circle representing a billion light-years. Now 14 billion years after the big bang, it shows the path of light emanating from Val toward earth. All galaxies are on the surface of the universe. This illustration shows the galaxies on this two dimensional time slice of the universe, including the Milky Way where Earth is located. The 4 arrows on the circumference depicts the fact that the circumference is growing at the speed of light.

The bottom illustration is the same time slice, but from the outside view where the universe is twelve dimensions, and every straight line is a helix. Hence, the size of this illustration has to be made 36 orders of magnitude larger than its actual size to be comparable in size to the top illustration. The bottom image would resemble a blurry image of the top illustration. It should be pointed out that the bottom image is how the universe actually is. The top illustration is just how it looks to us. Both contain the same three-dimensional volume. The bottom's volume is just spread over more dimensions.

To summarize: The universe started as a microscopic four-dimensional bag of atomic particles that fit easily into it. The bag expanded, eventually hugging the surface reaching its current, the speed of light. Gravity insured that all the particles would remain on the expanding surface, because it mostly acts orthogonal to the expansion. However, if some strayed away from the surface the majority pulled it back to the surface. During the 400 thousand years before atoms were formed, all the particles were in close enough contact to have equalized their temperature. Both of these factors kept matter uniformly distributed. The energy attributed to matter is kinetic, the universe expansion, which fuels all energy of the universe. The universe is finite, and just about totally visible. The expansion is not accelerating, but light from celestial object take curved path to us making it dimmer than expected. Once atoms were formed, slight density variations allowed matter to congeal into galaxies with large voids between them. Galaxies that formed less than about .6 billion years after the big bang can be seen from anywhere in the universe, and perhaps from more than one direction. The speed of matter is limited to the universe's expansion speed, but that requires understanding the 12 dimensional outside view of the universe, as previously explained.

And light is just a very small particle dispatched with a relatively large amount of energy, with its velocity limited to the speed of light. While dark matter is as compatible with this theory as with relativity, MOND is much more compatible with this theory, and is the preferred explanation. The bending of light by massive bodies as calculated in the treatise might also explain the bullet cluster apparent conflict with MOND. Interestingly, the same particle that causes gravity also causes the universe expansion, entanglement, and the strong force. But again, in order to explain it requires understanding the graveon mechanism, from the 12-dimensional hyper-helix view of the universe, also discussed earlier.

The Gyroverse
The Hidden Structure of the Universe

Donald Wortzman

PART II - CURRENT PHYSICS OVERVIEW

CHAPTER 8 - RELATIVITY OVERVIEW

INTRODUCTION

Sir Isaac Newton is the father of modern physics. His treatise Principia, completed in 1687, formulated his three basic laws of motion, which for most real-life problems still stand today. It has been said that he wrote his books in Latin, because he wanted to limit the commoners' ability to read it. If that were his reason, he greatly overestimated us. His invention of Calculus, simultaneously with Leibniz, is the most important tool of modern mathematics. It was needed to apply these laws of motion to astronomical problems and is now used for all scientific and engineering analysis. In addition, he discovered the formula for the gravitational attraction of two bodies and solved many longstanding problems of the motion of the solar system.

Newton was born around the time of Kepler and Galileo's death. Both Kepler and Galileo were the scientists who bridged the old and classical physics. They established the heliocentric model, where the sun was the center of the solar system, the planets revolve about it, and the moon revolves about Earth. It was Newton that gave physics its mathematical foundation and initiated it as a separate discipline of science. His three laws of motion and the gravitational law, summarized below are his major contributions to the physics of motion.

The first law of motion, borrowed from Galileo, the law of inertia, is that a body moving with a constant velocity will continue to move with the same constant velocity, unless an external force is applied to change it.

The second law of motion, the analytical form of the first law, is that the force is equal to the time rate of change of momentum, alternately stated as $F = d(P)/dt$ or $F = ma$.

The third law of motion states that every force has an equal and opposite opposing force.

Finally, he discovered all bodies attract each other with a force that is proportional to the product of their masses and inversely proportional to the square of their separation from each other.

Almost as important as his discoveries were the things he could not explain, but questioned. He wondered why the mass of inertia was the same as the mass of gravity. Having tried different material looking for a difference in these masses, he could find none. He was also puzzled by the immediateness of gravity as evidenced by the need of this assumption in calculating the correct planetary orbits. In addition, he thought that a gravitational attractive force must have an intervening medium.

Newton's theory remained for more than 200 years, not seriously challenged. A serious crack in the theory developed from an experiment by Michelson-Morley, in the late 1800's, which showed that the round trip speed of light was the same in all directions on Earth. To understand why this created a problem for the theory, some digression is necessary.

GALILEAN TRANSFORMATION

According to Newton, if an object is moving with velocity v_1 relative to a moving platform that is itself moving at velocity v_2 with respect to a rest platform, then the object is moving with a velocity of v_1+v_2. For example, if a river is running at 2 knots, and a boat is moving at 5 knots upstream would be moving at 3 knots relative to the shore. If the boat were going down stream instead, the speed relative to the shore would be 7 knots. This is very logical for a three-dimensional flat universe and correct to a high degree of precision. Surprisingly, this is not exactly true because the universe is more than three-dimensional. This will be discussed later in the chapter.

Physicists refer to this procedure of changing the reference between two inertial platforms, moving relative to each other, as the Galilean transformation or Galilean relativity, named in honor of Galileo Galilei. He came to this deduction by his recognition that the laws of motion seem invariant within a moving boat in calm water, or on land. Laws of nature that work on Earth, but are different on another planet, for example, would have a more limited application. Newton's laws of motion are covariant to this transformation, and work the same on any inertial platform.

The following equations describe the Galilean transformation in a mathematical formulation. For these equations the primed variables are the coordinates on an inertial platform that is moving at velocity v relative to the rest inertial platform with the unprimed coordinates. Also, v', v, and w are all moving in the plus x-direction.

Another way to view the situation is to imagine that the unprimed coordinates are on a space-platform near Earth, while the primed coordinates are on a spacecraft traveling at a high speed in the x-direction at velocity v. From the spacecraft, a smaller reconnaissance vehicle is heading further in the plus x-direction at velocity v' relative to the spacecraft. A monochromatic electromagnetic beacon on the spacecraft is beamed in the minus x (or minus x') direction toward Earth's space-platform, and another beamed from the space-platform to the spacecraft. For this situation, the set of transformation equations that convert events relative to the spacecraft (primed coordinates) to conditions relative to the space-platform (unprimed coordinates) are presented.

GALILEAN TRANSFORMATION CHART

Coordinates	$x = (x' + vt')$ $y = y'$ $z = z'$
Time Transformation	$t = t'$

	$\Delta t = \Delta t'$
Velocity Transformation	$w = (v'+v)$
Mass, Momentum & Energy	$m = m'_o$ $p = m'_o v$ $E = m'_o v^2/2$
Doppler Effect (Source Moving Away) **Doppler Effect** (Source Moving Toward)	$f = f'/(1+v/c)$ $T = T'(1+v/c)$ $f = f'/(1-v/c)$ $T = T'(1-v/c)$
Doppler Effect (Receiver Moving Away) **Doppler Effect** (Receiver Moving Toward)	$f = f'(1-v/c)$ $T = T'/(1-v/c)$ $f = f'(1+v/c)$ $T = T'/(1+v/c)$

COORDINATES

In the chart above, the first three equations express the coordinate changes between the two inertial frames. The z' and y' coordinates are unaltered because no movements in those directions take place. The spacecraft, the x'-coordinates, are moving away from the stationary space-platform in the plus x-direction. Its distance from the space-platform's x-origin, in time, is increasing by vt'. No loss of generality results in assuming all motion occurs in the x-direction, because the choice for the x-axis is completely arbitrary as long as the y-axis and z-axis are chosen in a way that makes all the axes orthogonal to each other.

TIME TRANSLATION

Since time is the same on both inertial frames, these two time translation equations transform time unaltered between inertial frames. For ordinary experiences, a timepiece on a moving platform keeps the same time that it would on a stationary platform. The

Galilean transformation assumes that time transforms unaltered even at higher speeds, outside ordinary experiences.

VELOCITY TRANSFORMATION

The reconnaissance vehicle relative to the space-platform becomes the sum of the velocities of the spacecraft with the reconnaissance vehicle relative to the spacecraft. Velocity is a vector number so that if the spacecraft and the reconnaissance vehicle are moving in opposite directions, one of the velocities would be negative and the result of the summation would be the difference of both quantities. Velocities v is that of the spacecraft relative to the space-platform, and v' is the reconnaissance vehicle relative to the spacecraft. The velocity of the reconnaissance vehicle relative to the space-platform is w.

MASS, MOMENTUM, AND ENERGY

These three quantities would use the same mass value, independent whether calculated on the space-platform or the spacecraft. In addition, mass is presumed to hold its value even for high speeds.

DOPPLER EFFECT

Doppler equations are dependent upon whether the light source or the receiver is moving. For most ordinary motion, the difference obtained with each set of equations is small. The first set of equations is for the moving spacecraft beaming toward the space-platform, and the second set is from the stationary space-platform toward the spacecraft. The frequency of light increases when the source and receiver approach each other and decreases when they recede from each other. Everyone is familiar with the siren of a passing police car. When it is approaching, the pitch is high, but then decreases as it passes. This is due to the Doppler frequency shift between a source coming toward and then going away.

GENERAL

These equations are accurate enough for most ordinary situations. The relativistic correction factor for these equations is in the order of magnitude of $(1-v^2/c^2)^{1/2}$. Even at high speeds, this relativistic correction is small. For example, given that light travels at $3*10^8$ meters per second, even for the fastest commercial airplane, the correction is about one part in a trillion.

LIGHTSPEED EXPERIMENTS

MAXWELL'S EQUATION

James Clerk Maxwell, 1831 to 1879, was one of the great physicists of his time. He was known for his ground-breaking work in electromagnetism and the kinetic theory of gases. After graduating with a degree in mathematics from Trinity College, Cambridge, he held professorships at several universities in England.

Maxwell's electromagnetic theory showed that equations could express the interrelated behavior of electric and magnetic fields produced by oscillating electric charges. In this theory, electromagnetism and the propagation of electromagnetic waves are expressed in four partial differential equations. Because the speed of propagation calculated from the equations was very close to the speed of light, Maxwell correctly proposed that light must be an electromagnetic wave.

In addition, since charges can oscillate with any frequency, Maxwell concluded that light was just a narrow band within a wide spectrum of electromagnetic waves. Included within the broader spectrum are radio waves, radar waves, microwaves, infrared radiation, ultraviolet rays, x-rays, and gamma rays.

AETHER (ETHER)

All waves in nature have a medium for traveling. It was assumed that electromagnetic waves also needed a medium in which to propagate. For example, sound can propagate in air or in other materials, but not in a vacuum. Associated with waves are mechanical vibrations that transmit the waves. Maxwell's equations, not covariant to the Galilean transformation, were considered further evidence that a medium for propagation was required. This perfectly elastic fluid for transmitting electromagnetic waves, which filled all the voids in the universe, was called aether. While no one was ever able to detect aether, it was thought to be present. It was assumed that Maxwell's equations describe motion in the inertial frame where the aether was at rest, and that Earth probably moved relative to this aether, since its direction of motion changes as it revolves about the sun. It was believed that one could calculate the speed of the aether by measuring the speed differential of light in different directions

MICHELSON-MORLEY EXPERIMENT

In the late 1800's, when the Michelson-Morley experiment was conducted, the technology did not exist to measure the speed of light directly. However, this experiment's purpose was to detect and measure the speed of Earth relative to the aether, and not determine the speed of light. The idea was to compare the round-trip speed of light in two perpendicular directions, using an interferometer developed by Michelson. This device was sensitive enough to measure this small difference in speed. They expected to measure a difference in the speed of light in the two directions, confirming the presence of the aether, but instead, to their surprise, the velocity of light did not vary. The following is a schematic of the Michelson-Morley experiment.

Figure 8.1 - The Michelson-Morley Experiment Setup

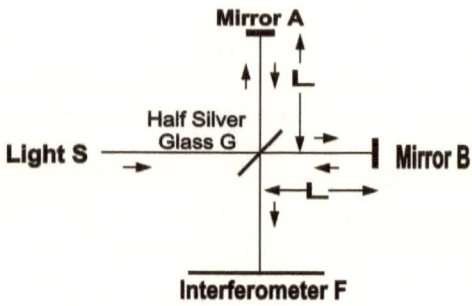

Figure_8.1 - This is a schematic of the Michelson-Morley Experiment. It was designed to compare the round trip speed of light in two orthogonal directions. For this example, assume that the aether is traveling in the up direction. The light originates at S, travels to the half-silvered glass G, which passes half the light straight through to mirror B, and the other half is reflected to mirror A. Both light beams bounce back to the half-silvered glass with half the returning light traveling to the Interferometer F. Though the time difference was in the order of magnitude 10^{-16}, the resulting diffraction pattern at F is enough to calculate the difference in time of travel for the two components of light, using the Michelson Interferometer.

According to the Galilean relativity, given e as the speed of aether, the speed of light in the direction of aether should be c+e and opposite the direction of aether is c-e. In both sideways directions, the speed of light should be $(c^2-e^2)^{1/2}$, since the small aether velocity in the vertical direction reduces the speed of light a small amount in the horizontal direction. Nevertheless, the round-trip time for the light to travel from the glass G to mirror A and back is still slightly longer than to travel from G to B and back by an amount (Le^2/c^3). The details of the calculation are not important. What is important is that according to the old Galilean relativity theory the time of light travel should have been different in both directions for this test. Then knowing the value of L and c would be sufficient to calculate e, the speed of the aether. The actual result was that the times for both round trips were the same, and consequently the speed of light in both directions was the same.

Since the direction of aether would not have been known, the experimental apparatus was rotated to maximize the difference. This, of course, did not change the results. The special relativity theory predicts that the one-way speed of light is the same in every direction, not just the round-trip speed. This test did not prove special relativity's more stringent postulate.

Thinking, perhaps, the relative velocity of the aether with respect to Earth just happened to be zero for this section of Earth's orbit, the experiment was repeated at different times of the year, when Earth was traveling in a different direction. This too did not make a difference.

The aether idea was not killed just yet because others had previously postulated that Earth being such a huge body could drag the aether as it passes through it. It would make the relative velocity between Earth and the aether zero, which would explain the Michelson-Morley results.

THE FIZEAU EXPERIMENT

The assumption that Earth could drag the aether conflicted with yet another experiment, conducted by Fizeau, which compared the speed of light in running water in the direction of flow and opposite to the flow.

Figure 8.2 - The Fizeau Experiment Test Setup

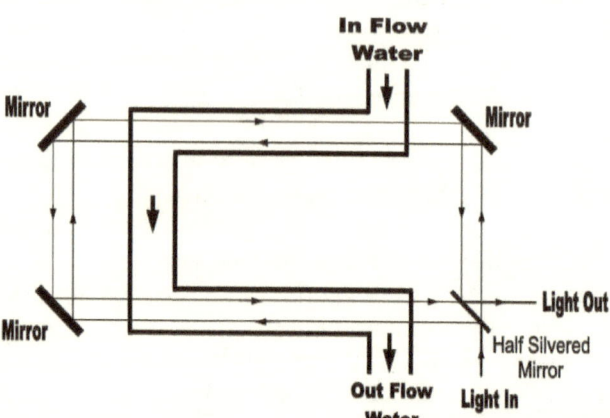

Figure_8.2 - Water flows in the conduit as shown. A beam of light is directed at the half-silvered mirror, from the bottom right. Half the beam is reflected and passes through the column of water in the bottom pipe, opposite to the flow of water. The beam is then reflected off the bottom left mirror then off the upper left mirror and through the upper column of water, again opposite to the flow of water. The beam is then reflected off the upper right mirror to the half-silvered mirror where half the beam is reflected arriving to Light Out. The other half of the beam from the light source passes through the half-silvered mirror and travels through the pipes opposite in direction, along the path shown, and rejoins the other beam to Light Out, where both beams are compared using an interferometer.

Outside the water, both beams travel an equal distance in air, although in different directions. Since light travels the same speed in all directions in air, the time of travel in air for both beams is the same. In the water, both beams also travel the same total distance, but one beam exclusively travels against the flow of water, and the other travels exclusively with the flow. In water, light travels three quarters the speed that it travels in a vacuum. As well, the speed is also dependent on the relative speed of the water. The resultant speed is not simply the addition of the two, as predicted by the

Galilean velocity transformation, but instead follows the Lorentz velocity transformation (to be described) formula. In other words, it follows the special relativity paradigm.

To explain the Fizeau results, the aether drag had to be something between full drag and no drag. While these were the highest profile experiments, other experiments also confirmed that the aether drag alone could not explain the consistency of lightspeed. Once again, this line of reasoning was abandoned, and a different explanation for the constant speed of light was pursued.

LORENTZ TRANSFORMATION

Hendrik Lorentz, born in 1853 in the Netherlands, was an outstanding scientist of his time. He was the last of the "old guard," a defender of the aether theory and classical mechanics, both replaced in the early twentieth century. He is most remembered for the Lorentz transformation and for his work with another great physicist from Ireland, George FitzGerald. They jointly developed the FitzGerald-Lorentz contraction theory. To get around the shortcomings of the orthodox aether theory, Lorentz and FitzGerald pursued a different line of reasoning.

In the year before Einstein presented his theory on special relativity, Lorentz proposed the Lorentz transformation. However, he first proposed it in 1899, but another physicist Joseph Larmor proposed it even two years earlier than that. This transformation would replace the Galilean transformation and have the characteristics necessary to explain the Michelson-Morley results. In effect, Maxwell's equations would be covariant to this transformation. For low velocities, it would reduce to the Galilean transformation, essential, since the Galilean transformation is logical and accurate for most ordinary motion. This transformation proposes that distances shrink in the direction of motion, and time measurements dilate (timepieces would slow) on inertial platforms in motion. The transformation explained most of the experimental results, by showing that distances would shrink in the direction of motion just enough to explain the constancy of the velocity of light in all directions. It also predicted a mass increase, besides time dilation, both needed to explain some experimental results. These phenomena were caused by the assumed effect that aether pressure would have on moving objects. Lorentz and others developed a non-convincing rationale to explain the cause of these effects. In effect, the equations were predictive, but the explanations appeared contrived.

Another important physicist involved in this debate was Poincaré. Henri Poincaré was born to a distinguished French family in 1854. He held chairs of mathematical physics and probability at Ecole Polytechnique, and the Sorbonne. His contributions were in many diverse areas in mathematics and the sciences, including the special theory of relativity. He articulated many principles of special relativity before Einstein, including that the laws of physics are covariant for all inertial frames, mass increase, and the non-simultaneity of time.

Perhaps Lorentz and Poincaré could not take the final leap toward special relativity because they could not accept the absence an explainable physical cause for the constancy of the speed of light in all inertial frames. Therefore, they would not completely let go of the aether theory. Interestingly, in a letter to Lorentz in November 1919, Einstein reversed himself somewhat on the existence of aether.

EINSTEIN'S AMAZING YEAR

At 26, Albert Einstein was an obscure technician working in the Swiss patent office in Bern. In the year 1905, the year after Lorentz presented his famous transformation equations; Einstein published five technical papers. Each paper recognized in its own right as a major contribution in physics, would have **etched his name in history.**

PHOTOELECTRIC EFFECT

One of these papers explained the photoelectric effect. When monochromatic light is shined on metal, all the electrons ejected have the same energy. If the light intensity is increased, more electrons are ejected, but still with the same energy. Physicists expected that as the intensity of light is increased, more electrons with greater energy would be dislodged. Einstein proposed that each electron is dislodged with a single photon. Each photon of a particular color (frequency) has the same energy, so that the electron is emitted with the energy over what is needed just to dislodge the electron. More intensity dislodges more electrons, but not with more energy. It takes a higher frequency light wave to dislodge electrons with more energy. For this paper, an underpinning of quantum mechanics, he belatedly was awarded the 1921 Nobel Prize in physics. However, he never won a Nobel Prize for relativity, his greatest accomplishment. While his relativity theory was universally accepted by then, it was not until 1923 that the Compton effect convincingly verified the major tenet of the photoelectric effect, so it is surprising that he won the award for that and not for relativity.

BROWNIAN MOTION

Another paper explained the cause of Brownian motion. Robert Brown, a botanist noticed that when small particles (pollen) are suspended in a liquid, they would pulsate randomly. Einstein explained that water molecules in the liquid bumping up against the particles caused this random motion. The consensus was that the millions of collisions on average would balance in all directions, and could not cause this motion. Einstein showed that over short periods, a great enough imbalance would occur to cause the random motion, and that he could calculate the average path of such particles. From this analysis, he also worked backwards to obtain a good estimate for the mass of molecules, not yet universally accepted to exist.

AVOGADRO'S NUMBER

This paper was for his PhD thesis. While it is lesser known, it gets the most annual citations today. It showed how to obtain Avogadro's number and the sizes of ions in solution from measurements of the coefficient of diffusion and its osmotic pressure. In 1811, Amedeo Avogadro conjectured that all gasses of equal volume and temperature contain the same number of molecules, called Avogadro's number.

SPECIAL RELATIVITY

That year, Einstein also published his most famous work on special relativity. This was spread over two papers. The first in effect, took the Lorentz transformation and gave it meaning. However, he re-derived the transformation starting with his two postulates of relativity. With his interpretation given to the transformation, he could glean much more intelligence from them. Lorentz and Poincaré needed to have an explanation for the constancy of the speed of light. Their intransigence was understandable, except no adequate explanation for the constant speed of light was forthcoming. Einstein on the other hand, accepted the evidence that the speed of light was a constant. Though the reason was not apparent to him, he investigated the consequences. The second paper probably presented the most recognized physics conjecture of all time, $E=mc^2$. It showed the equivalence of mass and energy. His special relativity theory was based on the following two postulates.

Postulate I - Principle of Relativity

All the laws of nature are the same in any inertial platform. The Galilean transformation had the same objective. Except, at the time it was proposed, fewer laws of nature were known. In particular, Maxwell's electromagnetic theory was not yet proposed. Einstein extended this principle to all the laws of physics, including electromagnetic waves.

Postulate II - Lightspeed Constant

This was accepting the results of the Michelson-Morley experiment as evidence for this peculiar quirk of nature, that the speed of light was a constant. Although the Michelson-Morley experiment only proved the round trip speed was the same in all directions, it was presumed that the one-way speed would also be the same. He entirely rejected the aether concept.

Additional Requirement

A third unwritten requirement was that the transformation had to default to the Galilean transformation for low velocities, all that is needed.

LORENTZ TRANSFORMATION - SR

Lorentz originally proposed the Lorentz transformation as part of the aether theory, but it was later adopted by Einstein to explain special relativity. However, a very significant difference in their interpretation applies. In the aether theory, the inertial frame where aether is stationary is the rest frame. Only in these frames, would the laws of nature not need a correction factor. Motion of objects relative to the aether would cause anomalies, for example, distance contraction.

In the special relativity interpretation of the transformation, any inertial frame can be considered the rest frame. Observers on the rest frame recognize length shortening on the moving frame in the direction of motion. Observers on the moving frame would not perceive this shortening of length, but would see similar length shortening on the rest frame. For them, their own moving inertial frame would be the rest frame, and the other rest frame would be the moving frame.

The following equations describe the special relativistic interpretation of Lorentz transformation. The primed variables are the coordinates on an inertial platform that is moving at velocity v relative to the rest inertial platform with the unprimed coordinates. Also, v', v, and w are all moving in the plus x-direction.

Similar to the Galilean transformation, a way to view the situation is to imagine that the unprimed coordinates are on a space-platform near Earth, with the primed coordinates on a spacecraft traveling at velocity v, in the plus x-direction, a large fraction of the speed of light. From the spacecraft, a smaller reconnaissance vehicle is heading further in the plus x' (or plus x) direction at velocity v' relative to the spacecraft. A monochromatic electromagnetic beacon on the spacecraft is beamed in the minus x' (or minus x) direction toward Earth's space-platform. For this situation, the set of transformation equations that convert events relative to the spacecraft (primed coordinates) to conditions relative to the space-platform (unprimed coordinates), are:

LORENTZ TRANSFORMATION CHART – S. R.

Coordinates	$x = (x' + vt')/\gamma$ $\Delta x = \Delta x'/\gamma$ $y = y'$ $z = z'$
Time Dilation and Non-Simultaneity	$t = \gamma(t' + [v/c^2]x')$ $\Delta t = \gamma(\Delta t' + [v/c^2]\Delta x'_e)$ $\Delta t = ([v/c^2]\Delta x_o)$
Velocity Transformation	$w = (v' + v) / (1 + [v'v/c^2])$
Mass, Momentum, and Energy	$m = \gamma m'_o$ $p = mv = \gamma(m'_o v)$ $E = mc^2 = \gamma(m'_o c^2) = \gamma E'_o$
Doppler Effect (Moving Away) **Doppler Effect** (Moving Toward)	$f = \gamma f'(1-v/c)$ $T = T'/\gamma(1-v/c)$ $f = \gamma f'(1+v/c)$ $T = T'/\gamma(1+v/c)$
Gamma	$\gamma = (1-v^2/c^2)^{-1/2}$

COORDINATES

The first three equations express the coordinate changes between the two inertial frames. The z' and y' coordinates are unaltered because no movements in those directions occur. The spacecraft, on the x'-coordinate, is moving away from the space-platform in the plus x-direction. Since the x-direction can be chosen arbitrarily, no loss of generality results by assuming the spacecraft is moving only in the x-direction. Its distance from the space-platform with time is vt'. In addition, because it is moving at the velocity **v**, all distance measurements in that direction are shortened by the factor $1/\gamma$, which is always less than one.

TIME DILATION AND NON-SIMULTANEITY

The second set of three equations is the time dilation and non-simultaneity equations relating time of an event on the moving inertial frame to the corresponding time of that event on the rest inertial-frame. The second terms in the first two equations are most surprising, since they show that time of an event also varies with distance. The second equation expresses the difference in time between two separated events $\Delta x'_e$ on the moving inertial frame. The third equation shows that two different places separated by Δx_o on the rest inertial frame will record different times for the same event on the moving inertial frame. Said differently, two separated observers on the same inertial platform would record completely different times for events on a moving platform. The third equation is obtained by setting $\Delta t'$ in the second equation equal to zero, and substituting ($\Delta x = \Delta x'/\gamma$) from the second equation in the first set of equations. Notice that this equation differs from the others in that the right side of the equation has the Δx_o term, a rest frame position measurement. Thus, the time of an event on the rest inertial frame varies both with place of the event on the moving inertial frame and the placement of the observer on the rest inertial frame. Even for low velocity, if the distances are large enough, the time adjustment can be significant. Remember that the speed of electromagnetic waves, the usual mode of event notification, is limited by the speed of light. These equations assume that the necessary corrections to account for the speed of light are made. It is not when we perceive events, but when they actually happen.

Although the difference in time between a pair of causal events in an inertial frame can vary when viewed from another inertial frame, the order of their occurrence will not change. This is because the difference in time between two causal events in a moving inertial frame is always more than the corresponding non-simultaneity relativistic shrinkage in time for the pair in the rest frame. In other words, the time change decrease between the two events will be less than the time it took for the first event to cause the second event, which is always limited by the speed of light. Therefore, the time between the events would not decrease enough to switch their order. On the other hand, if the causal pair bridges the moving frame and the rest frame, that is, not happening in the same inertial frame, it is not at all clear that the sequence of the events is always maintained. It will depend on how simultaneity is calculated. The orthodox interpretation would require that the calculation follow the procedure that maintains the order of events, but other seemingly correct procedures could reverse the order of events for some observers.

For example, consider a flyby shooting where someone in the moving frame shoots someone in the rest frame that he just passes. An observer in the rest frame next to the person shot would see the events in proper order. Consider two other observers on the rest frame, one far to the right and the other far to the left of the first observer. They will all agree on the time that the person was shot, since the observers and the person being shot are all on the same inertial frame. On the moving inertial frame the time the shot was fired would differ according to the three observers. The one downstream from the moving spacecraft would see it occurring at a time after the person was shot.

The most difficult aspect of special relativity to comprehend is the non-simultaneity term. To illustrate this point with another example, imagine another space-platform deep in space. Assume that this space-platform is in the same inertial plane as Earth space-platform. Further, assume just as the spacecraft reaches this space-platform out in space it decelerates abruptly and reverse course heading back to Earth at the same speed that it was formerly traveling away from Earth. Now consider what happens to the third equation in this set.

$$\Delta t = ([v/c^2]\Delta x_o)$$

For this example, the two observers that are being compared are at the deep-in-space space-platform and Earth's space-platform. Notice that when the spacecraft turns around, v goes from positive to negative. Observers in the deep-in-space space-platform, and in the craft will perceive the turnaround as immediate, but the observer on Earth's space-platform will view the turnaround as taking a long time, about $2\Delta t$ amount of time. For example, where the turn occurs 4 light-years away from Earth's space-platform, and ($v = .8c$):

$$(2\Delta t) = 2([v/c^2]\Delta x_o) = 2[.8c/c^2]4cyears = 6.4 \text{ years}$$

From the Earth space-platform observer's point of view, the turnaround took 6.4 years. Observers on either inertial frame near the turnaround will witness an immediate turn around.

TRANSITIVITY LAW & NON-SIMULTANEITY

A basic underpinning of almost all mathematics, including logic, which is just formalized common sense, is the Transitivity of Equality. In simple terms, what this principle states is that if $A = B$ and $B = C$, then $A = C$. This idea is so intuitive that it hard to imagine it not being correct. Virtually all mathematics proofs more than a few lines long, would invoke the transitivity principle at least once, and indeed, in the derivation of the equations in special relativity, it is used consistently throughout.

One consequence of non-simultaneity is that it violates transitivity, which leads to some bizarre results. This is perhaps the strangest consequence of special relativity, even more so than time dilation or distance contraction. To illustrate this, consider the train diagram that follows, Figure_8.3. Here we have two identical trains, the lower one is resting on the embankment track, and the top one is traveling to the right at high speed. Each car on this train has a timepiece. On the moving train, all the timepieces were synchronized among themselves such that all timepieces read the same. Since all the timepieces on the moving train are in the same inertial frame, this synchronization can be achieved. In fact, Einstein, himself, provided a simple procedure for doing it. Likewise, all the timepieces on the train at rest, on the embankment, can also be synchronized among themselves. The timepieces on the moving train would not be in synchronization with the timepieces on the rest train, because they are in relative motion. Each timepiece reading can be considered a distinct event. When time is viewed from the perspective of the moving train, the train's timepieces are in accord, but the rest train's timepieces are discordant. Assume the timepieces at A and U just happens to be displaying the same time when the cars are next to each other. This assumption simplifies the discussion but does not lead to a loss of generality. Timepieces B and V read one hour apart at this instant of (moving train) time. A real train would not be long enough, or fast enough for an hour time difference, but from a pure theoretical point of view, this is a perfectly acceptable assumption.

Figure 8.3 - The Proverbial Special Relativity Train

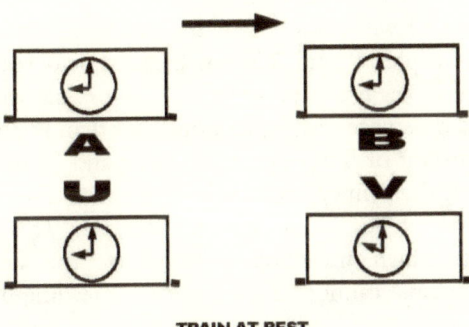

TRAIN AT REST

Figure_8.3 - The moving train timepieces are synchronized among themselves, reading the same time. Similarly, the rest train timepieces are also synchronized among themselves, reading the same time. However, when viewed from the perspective of the moving train, the moving timepieces read the same, but the rest timepieces on the embankment read one hour apart.

To synchronize timepieces on a common inertial plane, for example the moving train, a signal is sent from A to B. B turns around the signal immediately, sending back to A, but including his clock time. Since light travels at the same speed in any direction, both trips take the same time. When A receives the return signal with B's clock time, he adds half the time for the round trip to B's clock time to put his timepiece in sync with B's time.

According to people at A and U, both timepieces read just as shown, each at 9 o'clock. In theory, a fast camera could conceivably take a single picture showing both timepieces at 9 o'clock. The only persons who can attest that both timepieces read the same are people at locations A and U. At that instant, a person at A would attest to the fact that the timepiece at B also reads 9 o'clock, since both timepieces at A and B were synchronized. A person at B would concur since he also witnessed the synchronization procedure. Again, these are the only people in the universe that can attest to this firsthand. At that same instant, B would also say that V's timepiece read 10 o'clock. V would concur that when his timepiece read 10 o'clock, B's timepiece read 9 o'clock. V and B are the only firsthand witnesses who can attest to it. By invoking the transitivity principle, U and V's timepiece should read one hour apart, because U has the same time as A, and A has the same time as B, with U's timepiece reading an hour later than B's timepiece, all at the same instant of time. According to U and V, they can attest to the fact that their timepieces are in sync, both timepieces always read the same at the same instant of time. This leads to a contradiction, which must be resolved by the fact that the transitivity principle cannot be used to calculate time under special relativity.

Even more bizarre is that transitivity of equality was used to develop the relativistic equations. If equality is not transitive under special relativity, then it is questionable if it can be used to derive special relativity in the first place. Nevertheless, this leads to the paradoxes of non-simultaneity and of not maintaining the order of events. Of course, special relativity orthodoxy would not sanction this line of reasoning.

VELOCITY TRANSFORMATION

The next equation is the velocity transformation equation (velocity addition formula). It would be used, for example, to compute the velocity of the reconnaissance vehicle relative to the space-platform if the velocity of the reconnaissance vehicle was known relative to the spacecraft. Velocities v is that of the spacecraft relative to the space-platform, and v' is the reconnaissance vehicle relative to the spacecraft. The velocity of the reconnaissance vehicle relative to the space-platform is w.

The velocity of the reconnaissance vehicle relative to space-platform is always less than the v+v', the corresponding Galilean expression. Also, as v' approaches the speed of light, w would also approach the speed of light and would never exceed it. This equation is applied to explain the Fizeau experiment, where the passage of light was measured through moving water.

MASS, MOMENTUM, AND ENERGY

The next set of three equations is the mass, momentum, and energy equations respectively. Terms m'_o and E'_o refer to the mass and energy respectively, of an object in the moving inertial frame. These measurements are made in the moving inertial frame. This is a relativistic concept that matter not only has mass, but it can convert the mass entirely into energy, called the mass-energy equivalence. E, p, and m are the energy, momentum, and mass of the object as viewed from the unprimed or rest inertial frame. Energy E is the total energy of the moving matter, including its mass energy. The energy and momentum equations are shown in several equivalent forms.

The energy equation is not recognizable from its Galilean equivalent. There does not seem to be a kinetic energy term $(1/2 m'_o\, v^2)$. As the velocity approaches ordinary values, this term would result. To show this, the energy of matter factor has to be separated from the kinetic energy factor. By expanding gamma into an infinite series this kinetic energy term, $1/2 m'_o\, v^2$, can be separated out. This is derived in the Gamma Factor section that will appear later.

DOPPLER SHIFT

The Doppler expression relates the frequencies of the beamed electromagnetic wave between the space-platform and the spacecraft. Though the speed of light from any source measures the same in any inertial platform, the frequency of the light exhibits a Doppler shift when the source is from a different inertial frame. Since the platform and craft are receding from each other, the frequency is decreasing and the period of the wave, which is the reciprocal of the frequency, is increasing. The Doppler shift is only a function of the difference of velocity between the source and the receiver of the light. It does not matter which object is doing the moving. It is surprising that a Doppler shift occurs when the receiver is doing the moving. For example, if one moves away from a sound wave, the relative speed of the sound signal and receiver decreases, reducing the frequency of the received sound. Considering that the speed of light does not change at the receiver, even if the receiver moves, the frequency shift is surprising, and would not normally be expected. If, on the other hand, the source of the wave signal is moving away from the receiver, then a change in frequency is not surprising. A way to view this effect is that the beginning of each wave cycle starts at one location, but the end of the cycle happens either closer or further from the receiver, reaching the receiver either advanced or delayed.

GAMMA FACTOR

The gamma factor is the major factor for expressing the relationship between the resting and the moving inertial frames. The factor is symmetric, in that, each frame considers itself the rest inertial frame. Gamma is always one in the resting inertial frame and greater than one in the moving inertial frame. For a slow-moving inertial frame, gamma is close to one, and all the equations reduce to the Galilean equations. One equation where this procedure is not self-explanatory is in the energy relationship. To get the Galilean energy relation, the gamma term must be expanded into a series using the binomial theorem.

$$\gamma = (1 - v^2/c^2)^{-1/2}$$

$$\gamma = 1 - 1/2(v/c)^2 + 3/8(v/c)^4 + \ldots$$

For small v: $\gamma = 1 - 1/2(\mathbf{v}/c)^2$

Substituting into the energy relationship:

$$E = \gamma(m'_o c^2) = (1 - 1/2(v/c)^2)(m'_o c^2)$$

$$E = m'_o c^2 + 1/2(m'_o v^2)$$

$$E - m'_o c^2 = 1/2(m'_o v^2)$$

In Newtonian physics, nothing resembling mass-energy equivalence emerged.

Therefore, $m'_o c^2$ can be subtracted from both sides, resulting in:

$$K = 1/2(m'_o v^2)$$

where K is the recognizable kinetic energy.

GENERAL RELATIVITY

Mach's Principle

Ernst Mach was a physicist and philosopher, born in Austria in 1838, who died in 1916. He held various professorships of mathematics and of physics at several universities. He did much experimental work that was important in aeronautical design and the science of projectiles. His name has been given to a unit of velocity, Mach number, which is the ratio of speed of an object to the speed of sound.

He is also famous for Mach's principle, a name coined by Einstein. This principle postulates that inertia of any system is the result of the interaction of that system and the rest of the universe. In other words, all matter in the universe has a gravitational effect on all other matter, which causes inertia and centrifugal forces. This would explain why gravitational mass and inertial mass were the same. This, Mach could not quantify, since the gravitational force of far-off bodies is so minute with no discernible effect, that a case for it causing inertial forces cannot be analytically explained.

Newton took gravity, inertia, and centrifugal forces to be an innate property of matter. He was puzzled by the fact that inertial mass and gravitational mass were numerically equivalent. In fact, he conducted experiments to look for numerical differences between these two definitions of mass. Not finding any showed that the two masses have a common origin or resulted from an improbable accident of nature. He accepted their equivalence without ever understanding it.

GENERAL RELATIVITY VERIFICATION

Einstein, in his theory of general relativity, used the Mach's principle idea to help explain the equivalence of acceleration and gravity, which was dubbed the "Equivalence Principle." This principle, the cornerstone of general relativity, proposes that the effect of gravity in a local region is equivalent to the effect of acceleration of an inertial frame, without gravity. From this principle, general relativity extends special relativity limitation for dealing with constant velocity to situations with acceleration and gravitational forces. In principle, general relativity can address problems that are more complex. In practice the mathematics is much more difficult and even simple problems are hard to solve. Consequently, its value is more theoretical, understanding how nature works, than practical. The major discipline of mathematics required to solve general relativity problems is tensors, a generalization of vectors. An important consequence of general relativity is that very massive bodies distort space, and that gravity is caused by this distortion and not by a conventional force, as Newtonian physics teaches. For most problems, the correction offered by general relativity is too small to make a difference.

Advance of Mercury's Perihelion

One of the major initial successes of GR theory was to explain the larger than expected advance of Mercury's perihelion. The perihelion is Mercury's closest orbital point to the sun. Unlike the other planets that have relatively more stable closest points to the sun, Mercury's perihelion advances 43 arc seconds per century. In 1916, Einstein presented a paper covering the final form of general relativity, and correctly calculated the magnitude of this perihelion advance. Curiously, another physicist, Paul Gerber, obtained the same correct formulae 18 years earlier using a non-relativistic analysis. Einstein claimed this as a new result, although some of his nomenclature was reminiscent of Gerber's formulation. Gerber's work was later rejected for being obtained by a "false" method.

Another calculation, based on a different idea, was also made by Tom Van Flandern at Meta Research.

Also recently, Paul Marmet, a Canadian physicist from University of Ottawa, who died in 2005, calculated it resorting to nothing more than mass-energy conservation, and other classical extensions, some of which are also discussed in the LeSage section of the gyroverse theory. As a result, Marmet deduced the exact same equations as Einstein, for the advance of Mercury's perihelion. Marmet's calculation would also apply to the Gyroverse, and is endorsed by it.

While the Marmet analysis (Einstein's Theory of Relativity versus Classical Mechanics, Chapter 5) is the most convincing, he didn't have the benefit of the gyrohelix to explain the underlying mechanism involved. The energy and velocity of Mercury around the sun tilts the gyrohelix by gamma, which shortens its apparent path from rest space observation, making it longer on Mercury, and at the same time slows its travel time there. Also, Mercury's gravitational potential energy, which is twice that of its kinetic energy, tilts the gyrohelix twice as much. Together, they cause the Perihelion to advance its measured amount.

The important point is that there are several rationales for addressing the perihelion advance, and general relativity is not required for its calculation.

Bending of Starlight by the Sun

In 1911, Einstein's first prediction to verify general relativity was the bending of starlight as it passes near a massive object, like the sun. Bending was the same as the Newtonian model might have predicted for a light particle. Because it is not possible to see starlight while the sun is so bright, he suggested that the test be run during a solar eclipse. In 1912, efforts to test his prediction were prevented by cloudy weather at the observation site. In the 1916 paper, Einstein also discovered that his 1911 work was off by a factor of two, making the bending of light 1.745 seconds of arc for a star just grazing the sun. After another failed attempt, in 1919, Sir Arthur Eddington on an island off the coast of Africa and another group in Brazil confirmed Einstein's prediction. When announced, this prediction captured the imagination of the world and Einstein became an instant celebrity. Had the first attempt in 1912 to make this measurement not been called off by weather, Einstein's prediction then would have been wrong. One could only wonder what effect that might have had on Einstein's popularity.

CHAPTER 9 - QUANTUM THEORY OVERVIEW

EARLY HISTORY

INTRODUCTION

For over 2000 years, originating with Empedocles, the prevailing belief was that matter consisted of four elements: Earth, water, air, and fire. Each type of matter was made of different proportions of these four elements, giving it its unique characteristics. Aristotle, the great Greek philosopher born in Macedonia in 384 BC, refined this belief by adding a fifth element called aether, which he believed to be the main constituent of the celestial bodies. He believed light originated from the eye directed outward toward the objects, much as a blind man uses a cane. Some lone voices at the time thought that matter was made up of atoms, but none could offer adequate proof, so Aristotle's theory prevailed.

One of the first important challenges to this thinking was made by Abu Ali al-Hasan ibn al-Haytham, better known as Alhazen, born in 965 in Persia (now Iraq), who did much of his work in Cairo, Egypt. He gave the first correct explanation of vision, showing that light is reflected from an object into the eye. His explanation of light was unique for the time in that it was experimentally rooted, not based on abstract theory. Included was the formulation of reflection and refraction. The explanation is based on the idea that light is a particle that moves in straight lines, at a variable speed, being less in a dense material. He also described the use of a camera obscura, a crude version of the pinhole camera principle. In particular, the problem for which he is best remembered, namely Alhazen's problem, was finding the point on a spherical mirror where the light would be reflected to the eye of an observer.

LIGHT: A PARTICLE OR A WAVE

In 1690, Christiaan Huygens developed a theory that purported light to be a wave phenomenon. At the same time, Newton investigated the refraction of light by a glass prism into colors, and based his explanations of reflection and refraction on his corpuscular theory of light. It proposed that light was composed of tiny particles, or corpuscles, emitted by luminous bodies. Newton's corpuscular theory of light was favored over the wave theory. This thinking prevailed until Thomas Young and Augustin Fresnel did diffraction and interference experiments, which could only be explained from a wave theory of light. The four basic laws, known as Maxwell's equations, bridged electromagnetism with the theory of light, cementing the dominance of the wave hypothesis. Then, Einstein's photoelectric effect hypothesis fused the wave theory of light with Newton's corpuscular theory, originating the duality principle.

INITIAL ATOM THEORY

John Dalton, in 1808, made the first significant proposal of an atom theory of matter. Through his work on chemical compounds, he found that elements combine in very precise proportions of integral ratios of elements to form compounds. Because of this experimentation, he determined that the smallest indivisible particle of matter is the atom. He found that all atoms possess unique characteristics and weights. Unfortunately, his atomic proposal was not universally accepted at the time.

QUANTUM THEORY

Origin

Three names, whose discoveries in the early 1900s are associated with the origin of quantum theory, stand out. These three physicists are Max Planck, Albert Einstein, and Niels Bohr.

Max Planck was investigating black body radiation and discovered that electromagnetic waves emitted from objects were not distributed continuously, but emit energy in discrete quanta. This result, published in 1900, was the beginning of the evolution from classical mechanics to quantum mechanics. The quantum of action, now known as the Planck constant h, relates the energy of radiation to its frequency. For this work, he received a Nobel Prize in 1918.

The next quantum salvo, photoelectric effect, came from Albert Einstein in 1905, which built on Planck's results. This was the same year that Einstein presented his most famous work, the special theory of relativity, and some lesser-known works on the explanation of Brownian motion, and his PhD thesis on a method for obtaining Avogadro's number. Each of these works alone was significant, but the importance of relativity overshadowed the others.

Einstein's theory explained why shining monochromatic light on metal causes electrons with the same energy to be ejected. More intensity ejects more electrons, but with the same energy. At the time, it was expected that as the intensity increases, it would emit more electrons with greater energy. Einstein realized that each photon only dislodges a single electron. Each photon of a particular color has the same energy, so that the electron is knocked out with the energy difference of what is needed just to dislodge the electron. Increasing the intensity dislodges more electrons, but not with more energy. It takes a higher frequency light to dislodge electrons with more energy.

Niels Bohr founded the modern quantum theory of the atom. He was born on October 7, 1885, in Copenhagen, Denmark. Bohr studied at the University of Copenhagen, which he entered in 1903, and earned his doctorate there in 1911. He worked under British physicist, Ernest Rutherford, in Manchester, England. Rutherford's theory incorporated a nuclear model of the atom in which the atom is seen as a compact nucleus surrounded by a swarm of much lighter electrons. A problem with the model surfaced. According to classical physics, electrons orbiting the nucleus should lose energy until they spiral down into the center, collapsing the atom--this did not happen. Bohr took the quantum ideas of Planck and Einstein, combined with Rutherford's model, to develop his own quantum model of the atoms. In his model, electrons exist at set quantum levels of energy at fixed distances from the nucleus. The atom emits electromagnetic radiation only when an electron in the atom falls from one quantum level to another at a lower quantum level closer to the nucleus. If the atom absorbs energy, the electron jumps to a level further from the nucleus. His model was successful in making the theory fit the experimental evidence of the time. While over the proceeding years others ironed out inaccuracies in the model, the major idea remained. For this theory published in 1913 and 1915, he received the Nobel Prize in 1922. Einstein also received his Nobel that year, but his was presented belatedly for the previous year. Bohr commented that he would have felt awful if he got it before Einstein. They were fast friends, but at the same time were in competition as to the explanation of the non-local nature of quantum physics.

In general, measuring the electron energy states of these orbits directly is not possible. More commonly, every time an electron moves from a higher energy state to a lower one, it emits a photon whose energy is the difference between both energy states. By studying the emitted photon energies, it is possible to determine all the energy states.

While the Bohr model was a major step in understanding quantum physics, a chink surfaced, when better measurements found radiated frequencies the model hadn't predicted. These deficiencies were abated when it was realized that elliptic orbits and the magnetic moments would produce additional energy levels. Several different formulations were developed that explained this behavior. The first, a somewhat abstract theory called matrix mechanics, was developed by Werner Heisenberg. Following that, Erwin Schrödinger proposed an easier, less abstract formulation of quantum mechanics, known as wave mechanics. It emphasizes the continuous wave nature of matter as opposed to the matrix mechanics' discrete jumps. Soon after, Paul Dirac further unified both these formulations with his own theory. At this point quantum mechanics could explain the behavior of atomic structures by its three quantum numbers or so it was thought. Careful measurement of the energies of photons as electrons fell from higher energy states to lower ones indicated that the frequency of each transition was duplex, two frequencies very close to each other.

Up to this point, the discussion of the atom was confined to three quantum numbers, three of the four degrees of freedom that can be attributed to the electrons within the atom. As already described, the electron motion can be likened to the planetary orbits about the sun, with one major difference, electrons are restricted to only certain discrete energy levels. These quantum numbers are associated with; the distance that the electron is from the nucleus, the angular momentum of the electron as it orbits the nucleus, and finally the magnetic moment of the negatively charged orbiting electron.

In 1925, two graduate students at Leiden, George Uhlenbeck and Samuel Goudsmit, proposed the present accepted explanation of the fourth quantum state. They proposed that each electron spins with an angular momentum of one-half the Planck constant and carries a magnetic moment of one Bohr magneton. However, Paul Dirac, in 1928 provided the most elegant explanation, applying a relativity theory modification to Schrödinger's wave equation. It was applied to the hydrogen atom, describing the electron motion, including electron spin. Dirac's theory is relativistic in the sense that it is Lorentz transformation covariant. He published this work in 1930 in a book, <u>Principles of Quantum Mechanics</u>.

ELECTRON SPIN

Electrons exhibit a property of spin. Theory has it that the spin has an angular momentum of $+ h / 2$ and $- h / 2$. In other words, all electrons are spinning in one of two directions, up or down.

Because the electron is a charged particle, it also exhibits a magnetic moment corresponding to its spin and negative charge.

This characteristic of electron spin has some very troubling characteristics. It does not have any classical counterparts. It would be tempting to envision it as caused by the electron spinning around its own center with its negatively charged surface. Unfortunately, given the magnetic moment of the electron and the maximum estimate for the size of an electron, the electron's outer surface would have to be moving more than two orders of magnitude faster than the speed of light to have that much magnetic moment. Instead, spin is not considered actual rotation, but a pure quantum phenomenon that has no classical counterpart.

PARTICLE - WAVE DUALITY

The theory of light evolved from being thought of as a particle, through the influence of Newton, and back to a wave from Huygens theory and the experiments conducted by Young and Fresnel. The wave theory was further reinforced with Maxwell's electromagnetic theory that gave the wave theory a comprehensive analytical framework. This theory combined two forces of nature, electric field and magnetic field forces, into a single phenomenon. It also showed that light was just a narrow band of frequencies within the broader spectrum of electromagnetic waves. These waves include such things as radio waves, television waves, microwaves, radar waves, infrared radiation, ultraviolet rays, x-rays, and gamma rays. Just when the wave theory was universally accepted, Einstein explained the photoelectric effect by reintroducing light's particle qualities as a photon, a tight packet of waves. This evolved into the idea that light was both a particle and a wave, which interestingly, usually does not show both faces simultaneously. Experiments could be done that showed its wave properties, and other experiments that showed its particle properties.

The French physicist Louis de Broglie delivered his doctoral thesis at Paris University in 1924, where he extended the wave/particle duality concept to matter. He proposed that matter was also a wave where ($\lambda = h/p$) and ($\nu = E/h$). He calculated the wavelength of the electron in a hydrogen atom to explain the orbits of the electron in the hydrogen atom. The electron orbit spans an integral number of de Broglie wavelengths, generating a standing wave pattern that reinforces itself. This particle-wave duality was experimentally confirmed in 1927 by the Davisson-Germer Experiment, which obtained interference patterns from electrons passing through a crystal lattice. In 1929, de Broglie received the Nobel Prize in physics for this work. From 1932 to 1962, he was a professor of theoretical physics at the Faculté des Sciences at the Sorbonne

To complete the duality principle, Erwin Schrödinger, an Austrian physicist, formulated the theory of wave mechanics, which described the behavior of electrons in atoms in terms of waves, extending the de Broglie notion. For this achievement, he was awarded the 1933 Nobel Prize in physics with British physicist Paul Dirac who also made important advances in the theory of atomic structure. Schrödinger patterned his theory after the equations developed by Hamilton that was very successful in the analysis of classical mechanic problems. Schrödinger applied these equations to the Hydrogen atom, and got good agreement with Bohr's theory of the atom. Solution of these equations, instead of defining where the electrons are in the atom, gives the probability of finding them where they might be. The most probable solutions are referred to as the eigenvalues.

Duality applies to both light and electrons (and in fact all atomic particles), each can have both wave and particle properties. In addition, in any single experiment, only one of these dual faces is most often shown. This duality is sometimes explained by asserting that it is not the particle that is waving, but the probability wave function describing the particle that is generating interference patterns. Schrödinger himself never really accepted the idea that it was the probability that was waving, and not the particle.

One peculiar aspect of this duality is that many experiments start as probability waves and turn into particles only after being examined. For example, assume a setup having two possible outcomes, each with a probability of occurrence of 50%. As long as the result is not observed, the probability of the two outcomes remains intact, with each having a 50% chance of happening. Only after the result is examined, does the wave collapse to one of the two possible occurrences. One of the possibilities happens and the other does not. If this sounds strange, it also does to the most ardent supporters of the theory. Nevertheless, every time this aspect of the theory has been tested, it has been proven correct. It led Schrödinger to propose a "tongue-in-cheek" famous experiment to highlight this unbelievable consequence of quantum theory, which goes by the name, Schrödinger's cat.

SCHRÖDINGER'S CAT

In 1935, Erwin Schrödinger proposed a diabolical device to highlight one of the ways in which quantum mechanics contradicts our experiences of reality. In this thought-experiment, he imagined a cat in a sealed box, where poison would be activated by a radioactive particle that would kill the cat if it decayed. The radioactive material had a half-lifetime of one hour. A Geiger counter in the box was setup to release a hammer and shatter a small flask of poison gas, killing the cat. After an hour, the wave function would cause the probability of the outcome to be 50-50 for whether the cat was alive or dead. The probability of each outcome would change with time, but according to the theory, not until the box was opened and someone looked inside it would the wave function "collapse," and the probability change to certainty, rendering the cat either dead (retroactively) or alive. Until that time, the cat would be simultaneously both dead and alive with a certain probability of each.

What Schrödinger did was to combine an atomic model where we do not have any personal experiences and could accept indeterminacy, with a cat, where we have every day experiences to rely on, to show an absurd nature of quantum theory. It was not that he thought the theory wrong, but that it was incomplete, missing a vital part that would finesse its absurdness. As far as I know, no one has admitted performing this test. Many other tests have been run, and each time quantum theory was confirmed to high accuracy, leaving no doubt to its validity. The most important of these tests involved "quantum entanglement" and Bell's Theorem.

BELL'S THEOREM

ENTANGLED PARTICLES

Entangled particles are sister particles (usually photons or electrons) that share common origins and properties. For electrons, one has an up spin, while the other has a down spin. For photons, they might be polarized at right angles to each other or have the same polarization. Erwin Schrödinger pointed out that the theory allows a single quantum state, a particular polarization, for example, to be spread across two objects, such as a pair of simultaneously created photons, even if they are far from each other. He called this entanglement. The interesting thing about these particles is that they remain instantaneously (much faster than the speed of light) in touch with each other, no matter how wide the gap between them, even if separated to the extreme ends of the universe. It then predicts that a measurement on one particle will influence the outcome of a measurement on its distant sister, without any time delay. This phenomenon has been verified with photons to a distance of 7 miles.

EPR PARADOX

Einstein, notwithstanding his own major contributions to quantum mechanics, like Schrödinger, believed that quantum theory was not complete, missing some critical information. He referred to quantum entanglement as "spooky action at a distance." If we understood the missing pieces of quantum theory, he thought, all the spookiness would go away. To prove his point, he devised thought-experiments to show the absurdity of this aspect of the theory. Each time Niels Bohr, the most ardent supporter of the theory, would find faults with Einstein's argument, the volleying between them would continue. The last and most famous dialog was in a paper written by Einstein, and his colleagues, Podolsky and Rosen in 1935, now dubbed the EPR Paradox.

While Einstein, Podolsky, and Rosen originally envisioned using position and momentum, David Bohm, in his 1951 book <u>Quantum Theory</u> reformulated the EPR experiment using the total spin of the entangled particles, which must be zero. When the spin of a particle in one direction is known, according to quantum theory, knowing the spin in the other directions simultaneously is not possible. Overall, quantum theory limits simultaneous knowledge on any set of "mutually non-commuting observables." For example, it is not possible to know the position and the velocity of a particle simultaneously. The more precise one is known, the less exact the other. EPR devised a thought-experiment that would lead to a contradiction that showed that knowing all the directional spins simultaneously was possible.

For this experiment, they imagined sister entangled particles with opposite spin that were subsequently separated by a very large distance. When the spin of one of the particles is measured in the x-direction, the spin of the other in its x-direction is opposite. Orthodox theory claims it is not possible to simultaneously know the spin in the other two directions. Once any measurement is made on entangled particles, their joint wave function "collapses." This reveals the spin of both particles, which will be opposite each other in that direction, independent of the distance between the particles. EPR maintained that by choosing to make different axis spin measurements on one of the particles, in principle, they could have deduced the same axis spin measurements on the other particle. By placing the particles far from each other, relativistic speed limitation, would have prohibited timely communication between the particles. EPR surmised that the measurements could have only revealed preexisting values for the different axis spins of the second particle, contradicting quantum theory's dictum.

Bell's inequality and tests to implement it came much later, after both men were dead, and revealed that Bohr was correct. Quantum entanglement communication is non-local, not limited by the speed of light, so that even widely separated particles could communicate their spin instantaneously.

NICOLAS GISIN EXPERIMENT

Over the years, many tests of the entanglement phenomena have been done, each successive test fixing previous test flaws, but still leaving some wiggle room for the "Doubting Thomases." John S. Bell, in 1964, originally proposed the basic idea for this kind of test. Perhaps the most famous was the one done at the University of Paris by physicist Alain Aspect and his team in 1982. Recently, in 1998, Dr. Nicolas Gisin, of the Geneva University enhanced this test procedure, by extended the distance between the entangled photons from 100 yards to 7 miles.

In this test procedure, two entangled streams of photons were generated, separated into a right and left steam, and transmitted through separate fiber optic cables. They were passed through polarizer detectors that were identically aligned (at zero degrees), but 7 miles apart. Because each individual polarized photon pair was matched, and the polarization detectors were aligned alike, 100% of the two streams of photon pairs had matched polarization and tested as such.

Then the right polarization detector was rotated by plus 30 degrees. Because the difference in the polarization angle between the two streams was 30 degrees, quantum theory predicted that 25% of the photon pairs of the two photon streams would not have the matched polarization. This was confirmed by the experimental test results.

Next, the right polarization detector was set back to zero degrees and the left polarization detector was set to minus 30 degrees. Again the difference between the two polarization-detectors angles was 30 degrees, but in the opposite direction. Quantum theory again predicted that 25% of the photons of the two photon streams would not have matched polarization. This was also confirmed by the experiment.

Thus far, no conclusions could be drawn. For even if the two steams of photons were no longer entangled after separation, as both streams started matched, the results of both tests would have been the same. At this point one cannot be assured whether the results were due to the photon steams being entangled or just remembering their original state.

Finally, the right polarization detector was set to plus 30 degrees and the left polarization detector to minus 30 degrees. Then the difference angle of alignment between both polarization detectors was 60 degrees. Quantum theory predicted that if these particles were still entangled, 75% of the photons between the two streams should have unmatched polarization. On the other hand, if the two streams of photons were no longer entangled, but just remembered their previous polarization, normal probability theory would predict that at most 50% of the photon pairs should have unmatched polarization. This was because each stream separately only contributed a 25% difference in polarization. By comparing the average mismatch rate, one could decide with confidence whether the two streams were still entangled as they passed through the polarizing-filter 7 miles apart or if they were no longer entangled.

When this experiment was done, the mismatch rate turned out to be 75%, showing that the photon pairs were entangled to the very end. In addition, the polarizing filters were randomly switched "on the fly, just in time," such that no signal, even at the speed of light, could have warned each other of their polarization. This indicated that not only did they remain entangled, but also communication between them was much faster than the speed of light. In fact, the Schrödinger wave function predicted that this phenomenon was not just faster than the speed of light, but it was instantaneous. This experiment verified that the communication was at least 10 thousand times faster than the speed of light. Knowing that trillions, upon trillions, upon trillions of photons in a radius of 7 miles exist, it was an amazing feat, even neglecting the fact that it was also instantaneous.

In the Einstein pronouncement, he called this phenomenon "a spooky action at a distance." Einstein's EPR paradox conflict with Bohr that originally exposed this enigma, was not settled in their lifetime, but was finally settled many years later in Bohr's favor.

Amusingly, the current interpretation of special relativity argues that communication faster than the speed of light between entangled particles does not conflict with special relativity because it cannot be used to transmit messages faster than the speed of light. Apparently, Einstein thought it did pose a conflict since this was the crux of his argument in the EPR thought experiment. He used the lightspeed limitation to argue that the separated particles would not have sufficient time to communicate their entangled state. Interpreting this communication as a message would undermine the most basic tenet of special relativity. Revisionists felt compelled to narrow the definition of a message so as not to undermine the special relativity theory.

Why is this correlated communication between two particles different from a message from one to the other? It is probably only a matter of time before someone figures out how to use this concept for undisputed information transfer faster than the speed of light. It is possible that other civilizations are attempting to communicate with us through this type of mechanism, rather than electromagnetic waves that are too slow for communicating over significant inter-celestial distances.

An interesting variation would be to re-run Gisin's experiment using the same length of cable, but with the cables remaining closer together. By keeping the length of cables the same, the reduced effectiveness of the setup due to signal attenuation and most other sources could be kept constant. However, the shorter distance between the cables should increase the effectiveness of entanglement. Consequently, by extrapolation, the distance limitations of entanglement might be estimated.

THE STANDARD MODEL

The standard model, developed in the 70's, is the theory that explains all the interaction of the subatomic particles. It has successfully predicted the probable outcome of all particle experiments. Its limitations are that it does not address gravity, dark matter, dark energy, or account for the mass of the particles. It requires a multitude of unrelated ad-hoc parameters selected just to be consistent with observations.

The Atom

Nucleus

All matter consists of atoms that are about an angstrom (10^{-8} cm) in diameter. The major part of an atom is concentrated at the nucleus. The nucleus consists of protons, which are positively charged, and neutrons, which are neutral. Protons and neutrons, collectively called nucleons, are very small, approximately 10^{-13} cm in diameter, and occupy a very small part of the atom. Revolving around the nucleus are the negatively charged electrons, which are much smaller than the nucleons. The vast majority of the atom is empty space. Two atoms with the same number of protons, but a different number of neutrons, are called isotopes of the same element. The number of protons, which attract an equal number of electrons, determines the chemical properties of atoms. Elements with much more than 100 protons do not exist, because the electromagnetic repulsive force of the charged protons would be stronger than the attractive strong force that holds atoms together.

Electrons

Electrons, the negatively charged particles in an atom, can be simplistically viewed as revolving around the nucleus in fixed circular orbits called the energy orbits. Every electron has a charge of about -1.6×10^{-19} coulombs. The mass of an electron is 9.31×10^{-28} grams, about 1/1836 times the mass of a proton. Atoms nominally have as many electrons as protons, leaving the entire atom with a neutral charge. The number of electrons in the first four orbits is 2, 8, 18, 32, respectively, with the outermost orbit never having more than 8 electrons.

Elementary Particles

At the time that quantum theory was developed, it was thought that protons, neutrons, and electrons were all indivisible parts of the atom. During the 60's and 70's, it was learned that atomic particles could be broken down further. It was in vogue to give the newly discovered particles fanciful names. Thus, the "three generations" of elementary particles are assembled from six quarks (up, down, strange, charm, bottom, top), six leptons (electron, muon, tau, and a neutrino associated with each of the three), and four types of force-carrying particles, the bosons (photon, gluons, graveon, vector bosons Z and W). Only particles in the first generation are found in atoms. The other generation of particles are duplicates whose function has still not been identified. They are rare and hard to produce, usually requiring high-powered particle accelerators. Their importance seems mostly centered in developing the theory behind the particles. Protons are formed from three quarks, two up quarks, and one down quark. The charge of these quarks total up to one (2/3 + 2/3 - 1/3). Neutrons, on the other hand, are also formed from three quarks, but two down quarks and one up quark. The charge in this case is zero (2/3 -1/3 - 1/3). Electrons are the only lepton found in an atom. All the bosons, the force carrying particles, are found in atoms, and together are responsible for the four forces in nature. In total, hundreds of particles are known. There are actually many more quarks, since they also come in "colors and flavors." In addition, the plethora of particles includes antiparticles that produce antimatter.

Fermions and Bosons

Another classification for particles is fermions and bosons. Quarks and leptons together comprise fermions, having half integer spin (1/2, 1 1/2, etc.) and obeying something called Fermi-Dirac statistics. They have a solitary property resisting compression toward each other, which leads to the Pauli exclusion principle; no two can occupy the same state simultaneously. All chemical bonding and the structure of matter are a result of this action. This keeps matter from compressing. Bosons, the force carriers, on the other hand, gather and may overlap in the same quantum state. In fact, the more bosons that are in a state, the more likely that still more will join. These kinds of particles can pack tightly together. Bosons have a full integral spin (1, 2, and so on.) and follow the so-called Bose-Einstein statistics.

Four Forces of Nature

Gravity

Gravity is the weakest of the forces, 10^{-40} times the strong force, and mediated by exchanging graveons, which has a rest mass of zero and spin of 2. It is responsible for keeping the universe and our solar system in place, the sun in the Milky Way, the planets revolving around the sun, and the moon revolving around Earth. It keeps us and all our possessions bound to Earth. The fact that it acts between all matter, and is only attractive, dropping off slowly by the square of distance between matter, gives it a very big effect, though next to the other forces it is weak.

Newton depicted gravity as an immediate force, with no time delay. General relativity describes gravity as a bending of space-time, and administered by gravity waves that travel at the speed of light. String theory reintroduces gravity as a force, mediated by the graveon, attempting to unify it with the other forces.

Electromagnetism Forces

Once, these were considered two different forces, electric and magnetic. Maxwell's equation brought them together. It is mediated by exchanging photons, which has a spin of one. This force also acts long range like gravity and drops off at the same rate. It is attractive for opposite-charged particles and repulsive for like-charged particles. Although it is much stronger than gravity, 10^{-2} times the strong force, it is not very far-reaching. This is due to the net charge on most large objects being zero, which neutralizes the force. A piece of paper held down by all of Earth's gravity can be picked up by the charge on a comb, overcoming gravity. In the atom, it is responsible for the chemical bond between elements to form compounds. It is also the mechanism behind such things as mirrors, electric motors, and magnetic compasses. It is the reason that one's hand does not go through something touched.

Weak Force

Two kinds of weak interactions are mediated by exchanging one of the two W or Z bosons. The weak force is responsible beta decay, which governs the decay of neutrons, protons, and many other particles. It is a very short-range interaction, restricted to distances shorter than the atomic nucleus, 10^{-15} centimeters. Its strength is about 10^{-8} times the strong force. Physicists have recently explained that the weak force and electromagnetic force are fundamentally the same force, called the "electroweak force."

Strong Force

The most powerful of the forces, appropriately known as the strong or nuclear force, binds together the protons and neutrons (that is nucleons) that comprise the atom's nucleus. The force is mediated by particles called gluons, a massless, chargeless, spin 1 boson that are constantly exchanged between the nucleons. The strong force has a very short interaction distance, limited to within the nucleus, 10^{-13} centimeters. The strength of the force is two orders of magnitude greater than the electromagnetic force, eight orders greater than the weak force, and forty orders greater than gravity.

Quantum Electrodynamics (QED)

Quantum electrodynamics, called QED, is a joining of quantum theory with electromagnetism. It underlies the interaction between photons and electrons. In 1965 Richard Feynman, Julian Schwinger, and Sin Itiro Tomonago shared the Nobel Prize in physics for their contributions to electrodynamics. Charged particles interact by emitting and absorbing photons, the particles of light that transmit electromagnetic forces. For this reason, QED is also known as the quantum theory of light. Feynman developed a way to analyze these problems using a system of notation called Feynman Diagrams. QED events are broken down into three basic actions to produce all the phenomena associated with light and electrons:

1. Electrons moving from place to place.
2. Photons moving from place to place.
3. Electrons absorbing or emitting photons.

The combination of these simple events describes most familiar everyday phenomena. For example, all chemical reactions and all biological processes can be understood by these interactions. The characteristics of solids and liquids can also be described by it. Even mirrors are explained by these interactions. The two notable exceptions are gravity and nuclear reactions. Feynman created his diagrams as a notation for these events. In a Feynman Diagram, photons are squiggly lines, electrons are solid lines, and anything moving at the speed of light, c, is represented by a 45-degree angle. His method of analyzing quantum problems is fast, accurate, and illustrative of the process.

Quantum Chromodynamics (QCD)

A cornerstone of the standard model is quantum chromodynamics (QCD), which describes the interaction between quarks and gluons in the atom's nucleus. As the acronym suggests, this work was pattered after QED. Despite the positive charge in the nucleus of atoms, it stays together because of an attractive force, called the strong nuclear force, which is stronger than the electromagnetic repulsive force. The person most associated with this theory is Murray Gell-Mann, who received the Nobel Prize in Physics in 1969. He proposed a new quantum property of particles, "strangeness," and sorted particles into eight "families," named for a Buddhist philosophy. This work led to the postulation of the quark particles, the assumed constituents of protons and neutrons, whose name was whimsically adopted from James Joyce's "Finnegans Wake."

QCD produces some of the most computationally intensive problems known to the physics community. Even with today's powerful high-performance computers, scientists must develop algorithms that efficiently use a machine's capabilities to study this theory.

Electroweak Force

Three people most associated with this theory are Steven Weinberg, Sheldon Glashow, and Abdus Salam. They shared a Nobel Prize in 1979 for uniting these two forces into a single force known as the electroweak force. Soon after the big bang, the electroweak force separated into two forces by a process referred to as symmetry breaking. This occurred as the universe cooled an infinitesimal fraction of a second after its creation.

Higgs Field and Higgs Boson

A deficiency of physics is that it does not predict the mass of the fundamental particles. Peter Higgs, at the University of

Edinburgh in Scotland, in 1964 proposed a partial solution. He hypothesized that all space in filled with an undetectable field, and an associated particle, (dubbed the Higgs field), which gives all matter its mass. According to this hypothesis, the amount of drag a particle receives attempting to change motion in the field gives rise to its inertial mass. Recently, a particle presumed to be the Higgs has been found, but there is no physical evidence that it performs the functions expected, for example, creates a particle's mass.

STRING THEORY

INTRODUCTION

The standard model of Elementary Particles describes how the elementary particles are organized and how they interact with one another via the forces of nature. While the standard model has been very successful in describing most of the elementary particle phenomenon, it leaves many unanswered questions about the fundamental nature of the universe. The goal of modern theoretical physics has been to find a unified description of the universe, a single theory to explain all the atomic particles and the four forces of nature, including gravity.

Unification has historically been a very fruitful approach. For example, Maxwell unified the forces of electricity and magnetism into the electromagnetic force. The Nobel Prize winning work of Glashow, Salam, and Weinberg successfully showed that the electromagnetic and weak forces could be unified into a single electroweak force. Einstein showed that mechanics and electromagnetic theory are covariant to a Lorentz transformation. In the latter part of his life, Einstein tried without success to unify relativity and electromagnetic theory.

GUNNAR NORDSTRÖM THEORY

A first attempt to unify all the forces of nature was done by Gunnar Nordström in 1914. He taught at the University of Helsinki, and later in 1918 became professor at the Helsinki

University of Technology. For his unification theory, he assumed the existence of a fourth space dimensions, and was able to combine electromagnetism with gravity. The output was Maxwell's electromagnetic wave equations together with the equations of Einstein's special relativity. At that time the strong and weak forces was not yet known, so his theory combined all the known forces. His theory later evolved into a competitor of general relativity in explaining gravity, but was less predictive than it, although it was better than Newtonian theory in explaining several celestial anomalies.

KALUZA-KLEIN THEORY

Another important attempt to unify the forces was made by Theodor Kaluza in 1921. Kaluza was teaching at Königsberg in April 1919, when he wrote to Einstein about his ideas to unify general relativity and Maxwell's theory of light. Einstein encouraged him to publish his work, and he did so in 1921, his only paper on the subject. In the original work of Kaluza, perhaps borrowing from Nordström, it was shown that starting with a theory of general relativity in five space-time dimensions, a four-dimensional theory of Einstein's general relativity plus Maxwell's electromagnetic theory would result.

Despite his early encouragement and recognition of Kaluza's good idea, it seemed that Einstein was never convinced of the efficacy of the five-dimensional hypothesis. He ultimately did not support the combination of electromagnetism and gravitation differently than discussed in his own 1922 book, <u>The Meaning of Relativity</u>. However, in later years, to no avail, he dabbled with a fifth dimension, in his quest to develop a unified theory.

Oskar Klein made an important modification to the five-dimensional theory in 1926. He independently came up with the same idea, but explained how the fifth dimension could be rolled up to the size of 10^{-33} cm Planck length, and not be seen because it is smaller than any particle. He postulated that the extra dimensions are physically real, and that they could lead to the observed forces beyond gravity when looked at from our four-dimensional world.

For many years, Klein attempted, in different ways, to promote the fifth dimension as a convenient way to introduce the quantum of action into the gravitational field.

Due to all the work going on in quantum mechanics, interest in the Kaluza-Klein theory waned. Interestingly, the theory brought two powerful notions to physics. First, was that the universe might have more than three space dimensions. Second, it proposed that the extra dimensions might not be seen, because they were exceedingly small and tightly rolled up out of view. These two ideas were the needed seeds for developing a unification theory, known as string theory. However, because the gyroverse theory contains a well-defined extra space dimension in its Euclidean view, it would make sense to revisit the Nordström and Kaluza-Klein Ideas, especially since time in GR when multiplied by c is the analog of the gyroverse's expansion at the speed of light. This could be a stepping-stone to the development of the gyroverse theory.

STRING THEORY AND M-THEORY

While string theory was developed by many people, it died and was resurrected several times from the 70's to the 90's. String theory is an attempt to create a theory-of-everything (TOE) that unifies the electroweak, strong nuclear, and gravity forces. As mentioned, the electromagnetic and the weak nuclear forces were unified earlier in the 1970's.

The earliest names associated with string theory are John Schwarz of Caltech and Michael Green of Cambridge, working together in the 1980's. The basis of the theory is that all matter is composed of strings. Strings are the smallest possible object, with a length of 10^{-33} cm, and no width or height. In current thinking, fundamental particles are thought of as point-like, zero-dimensional objects. Strings are larger, but still very small as compared with the length scales that can reasonably be measured. For most practical cases, they behave as point particles. This stringy nature has important implications. They are one-dimensional and either can be open or closed. Closed strings have the shape of a circle or oval, while open strings have ends. Superstring theory is string theory that also relates bosons and fermions to each other.

Strings are free to vibrate, and different vibration modes of the string represent the different particle types, different masses, and spins. One mode of vibration makes the string appear as an electron, another as a photon. The importance of this is that in a way it reduces the number of elementary particles to one. It shows that all particles are constructed from the same string object. One mode even describes the graveon, the particle that produces the gravitational force. It is another important reason string theory has received so much attention. Two shortcomings of the theory were that it only worked in 10 or 26 dimensions, and that five different variations emerged.

Mathematical anomalies appear for any other number of dimensions. This begs the question, "where are the other six dimensions?" since we are only cognizant of four of them. The Kaluza-Klein theory proposed that it is possible for a dimension to be curled up into an extremely tiny ball 10^{-33} cm long, which is not detectable. For string theory, this occurs on a mathematical elaboration known as Calabi-Yau Manifolds. It is theorized that this is what has happened to the other six dimensions after the big bang.

Having five string theories was also a conundrum. Recently, Ed Witten, professor at the Institute for Advanced Study in Princeton, postulated that the five string theories are actually different views of the same overriding theory, M-theory, not different theories. Because of all his contributions, he is widely recognized as the foremost mathematical physicist of our time, receiving the Fields Medal in 1990, mathematics Nobel equivalent. M-theory adds another dimension to string theory, making the total, eleven dimensions.

Since then, the five string theories have mushroomed by over 100 orders of magnitude. Some physicists have proposed that each of them has lead to a different universe, which is part of a greater multiverse. Our universe is unique among them, because it has the parameters that spawned observers that are aware of it. This idea known as the Anthropic Principle was proposed by Leonard Susskind of Stanford University.

While string theory might be helpful in unifying physics, it falls short by not making any significant new verifiable predictions.

Making verifiable new predictions is the passport to a theory's credibility. Several well-respected string theorists, Lee Smolin of the University of Waterloo and Peter Woit of Columbia University have written books critical of the theory. Their biggest concern is that almost all research on physics unification is limited to string theory, neglecting possible alternate investigations. Considering the paltry success to-date, this monolithic approach on the part of the physics community is not justified.

SYMMETRY AND GROUPS

The more physicists examine the physical laws and the particles of nature, the more apparent are the special relationships they have with each other. Symmetry and groups are nature's way of dispensing these common relationships. Studying the patterns that emerge enhances the understanding of the universe.

Examples of symmetry are translation and rotation. When moving to a different location or turning, the same laws of nature apply. Another symmetry is the invariance of the physical laws to uniform motion. Inside a smoothly cruising plane, one is not aware of the speed, because the laws of nature are not different in it. The first encounter with this notion was the Galilean transformation, which was symmetric with respect to the physical laws, but not symmetric with respect to electromagnetic waves. Einstein recognized that nature had a propensity for symmetry and proposed the Lorentz transformation because it made it symmetric to both.

A very enlightening theory regarding symmetry was discovered by Emma Noether, who was born in Germany in 1882, and is considered by many as the greatest women mathematician of all time. She was an unofficial professor at Göttingen, then one of the most prestigious universities of mathematics in the world. It was unofficial, because women could not hold such professorships in Germany at that time. Consequently, she was not paid. David Hilbert, another great mathematician had allowed Noether to lecture her courses under his name, until she was officially placed on the Faculty. Noether discovered that each symmetry produces a conservation law. Translational symmetry, where the laws of nature are independent of location, causes the conservation of momentum. Time symmetry, where the laws of nature are invariant with time, causes the conservation of energy. It is interesting that these physical principles of import are not just arbitrary accidents of nature, but are derivable from first principles.

A group is a collection of objects or abstractions with a common operation or rule that transforms the elements into each other. Evariste Galois, born in France in 1811, and killed in a duel when he was just 21 years old, invented the mathematical notion of groups. The legend, surrounding this duel, was that it was over an unrequited love, and that the night before, he wrote out his theory of groups. It is amazing that that a class lasts for a year what he wrote in an evening. A special continuous sub class of groups is known as Lie groups. These groups have geometrical representations that are closely associated with particle physics.

All the integers, both negative and positive, and the addition operator make up a group. If any two integers are taken and added, another integer results. A more abstract example would be a 90-degree rotation group. This group has four members, one that rotates 90 degrees, another that rotates 180 degrees, still another member that rotates 270 degrees, and finally an identity element that leaves everything unchanged because it rotates 360 degrees. By applying any two of these in a row, results in another member of the group. For example, the rotation of 90 degrees, followed by a rotation of 180 degrees is the same as the 270-degree rotation. A 270-degree rotation, followed by a 180-degree rotation is the same as a 90-degree rotation.

It turns out that the atomic particles bare relationships to each other, albeit more complex, as group members. Physicists have uncovered these groups and found new member particles of the groups that were not known beforehand. They predicted their existence and characteristics by their placement within the group, demonstrating the usefulness of the group concept.

Donald Wortzman

CHAPTER 10- COSMOLOGY OVERVIEW

ANCIENT COSMOLOGY

Greek astronomers devised the basic astronomy ideas and theories until Copernicus. The central theme of ancient Greek astronomy was the geocentric theory that claimed that Earth was at rest and the center for all other motion in the universe. The Greeks, who did not know about gravity, thought that if Earth were moving, objects not attached to Earth would not move with it.

Ancient Greeks divided the universe into two domains, the celestial domain that included the sun, moon, stars, and planets, and the terrestrial domain that included Earth and its atmosphere. The celestial domain was non-changing and embedded in transparent spheres that encircled Earth. The terrestrial domain did not move.

Many ancient Greek astronomers made important contributed to this science. One of the most prominent of these was Aristotle who lived between 384-322 BC. He was the philosopher's philosopher in that his thinking dominated much of intellectual thinking for two-millennium. He believed that the sun was a sphere, and more distant than the moon. By observing the lack of stellar parallax, which is simply the expected annual change in the observed relative positions of the stars from Earth, he concluded that Earth must have not been moving. However, due to the immense distance to even the closest stars, stellar parallax needed very accurate observations with a telescope to be detected. In fact, stellar parallax was only first observed in 1838.

Another prominent Greek astronomer was Ptolemy, who lived in Alexandria, a primary center of Greek culture between 85-165

CE. Little was known about his life other than his considerable writings on a variety of subjects. He was most noted for extending Hipparchus's system of epicycles to explain his geocentric theory so that it would better account for the observed motions of the planets. Since the planets really revolve around the sun, from Earth, the planets appear to be moving backward at times. If it were thought that the planets rotate around Earth, these smaller circular, or epicycles, were needed to explain this apparent retrograde (backward) motion. He proposed that instead of the stars revolving in circular orbits around Earth, the stars revolved around a point called a deferent. The deferent revolved in a circular path around Earth. In this way the stars then traveled in a circular orbit around this new deferent point, which was offset from Earth's center. This motion enabled him to resolve two problems with the original theory. The first was that some planets, thought to be stars, seemed to follow a circular orbit about a point not centered in Earth, but about a point shifted from Earth slightly. The other problem was a discrepancy in previous explanations of the retrograde motion of planets. This theory explained the three motions of the planets, known as the Ptolemaic system, predicting the positions of the planets accurately enough for naked-eye observations. They consist of:

Diurnal: This represents the most prominent heavenly motion. It is the daily westward motion of the celestial sphere thought to contain all the stars fixed to it. It reflects the rising and later setting of the stars, which is caused by Earth's eastward rotation.

Prograde: This effect is the second most prominent motion. It is the slow eastward movement of planets along the ecliptic. The ecliptic is the plane of motion of all the planets in the Solar System. The planets orbit the sun at different angular velocities but remain predominately in the same two-dimensional plane.

Retrograde: This effect is more subtle than the others. It is a temporary westward (backward) motion making the planets appear to travel in looping, backward paths along the ecliptic. This effect comes about as Earth passes other planets that are orbiting the sun slower than Earth. Because Earth is closest to the planets at this point, the planets appear to be moving backward relative to the other stars in the sky as Earth passes it. This lasts for a few weeks until the planet resumes its forward motion.

The Ptolemaic system was universally accepted until Copernicus' heliocentric theory (sun centered), some 1600 years later, was proposed. This marked the end of the influence of the ancient Greek astronomers and their geocentric era. Fascinatingly, the ancient Greek astronomer, Aristarchus, did have similar ideas to those of Copernicus, but they were rejected at the time. Any idea that contradicted Aristotle and Ptolemy had little chance of being accepted.

CLASSICAL COSMOLOGY

NICOLAS COPERNICUS

Classical cosmology began with Nicolas Copernicus, a Polish astronomer who lived between 1473-1543. He studied mathematics and optics at Cracow University in Poland, and then studied canon law in Italy. While at the cathedral of Frauenburg, he developed his theory. Copernicus degraded the importance of Earth when he introduced the heliocentric model. This model had the sun at the center of the universe and Earth revolving around it each year while rotating once every twenty-four hours about its own axis. He claimed that all the planets, not only Earth, moved in orbits around the sun. Then he showed how this new system could accurately calculate the positions of the planets.

Copernicus spent many years developing the mathematical proofs for heliocentric orbits. While the math was less complex than that of Ptolemy, it still relied on multi-circular orbits to describe the motion of the planetary bodies accurately because he assumed that all the planetary motions were circular. Copernicus was reluctant to publish his work, not so much because of concern with repercussions from the church, but rather because he was a perfectionist. He never thought the work was complete, even after 30 years. On his deathbed, he finally did see his great work, <u>De Revolutionibus</u>, in print. Copernicus' original manuscript, and all its revisions, lost to the world for 300 years, was found in Prague in the middle of the 19th century.

For a long time, both Copernican and Ptolemaic theories stood side by side. Copernicus' model was more useful for calculations, but Ptolemy's theory was more palatable and consistent with human experiences on Earth. Earth did not seem to be moving, and no theory could account for the relatively stable structure of the planet as it revolved around the sun. This pluralism lasted until Galileo and Kepler found better evidence for the heliocentric model.

GALILEO

Although Galileo studied medicine at the University of Pisa, he was mostly interested in mathematics and science. After teaching mathematics at the University of Pisa, he was appointed mathematics professor at the University of Padua, in the Republic of Venice. He is mostly remembered for his efforts on the laws of motion, discoveries by his use of the telescope, and the pioneering of modern experimental science.

In 1609, having heard of a magnifying instrument developed by a Holland lens-grinder, he constructed the first complete astronomical telescope. Galileo used this 30-power telescope to discover moon craters, sun spots that rotated with the sun, Jupiter's largest moons, and phases of Venus. This last observation showed that the Copernican heliocentric theory was correct, since phases could only be observed if Venus was closer to the sun than Earth, so that a partial backside view could be seen of Venus, with its crescent shape. In addition, the size change of Venus between its closest and furthest point from Earth caused vast difference in its brightness, as viewed from Earth. Galileo sited these phenomena as evidence that Venus really revolved around the sun, not Earth. The moons of Jupiter also gave credence to the fact that all heavenly bodies do not necessarily revolve around Earth. In 1611, Galileo published his observations in <u>Siderius Nuncius</u> (<u>The Starry Messenger</u>). In the final years of his life, he was placed under house arrest as a heretic, only to be posthumously exonerated by the pope in 1992.

Kepler

Johannes Kepler was a German astronomer, born in 1571, the son of a poor mercenary soldier. He received his education through the scholarship system in Württemberg, which was set up to produce teachers and Lutheran pastors. In 1589, Kepler entered the theological seminary at the University of Tübingen, where he learned Copernican astronomy, and was awarded an MA in 1591. He had planned to pursue a religious life, but instead accepted a chair in mathematics and astronomy at Graz.

Kepler, a more theoretical contemporary of Galileo, wanted to analyze the motion of the planets. He moved to Prague to work with Tycho Brahe, the well-known Danish astronomer. After Brahe's death, he assumed his position as Imperial Mathematician. Starting with Brahe's extensive data, compiled years earlier at his former observatory near Copenhagen, Kepler found patterns in the movements of the planets, and deduced three laws of planetary motion.

The first was that planets move in an elliptical orbit around the sun, with the sun at one focus. Copernicus had assumed that the planets moved in circular orbits around the sun. This was a good approximation, but still relied on multi-circular orbit models to accurately describe the motion of the planetary bodies. The correction was needed to compensate for error in orbital motion introduced by the assumption that the planetary orbits were circles.

Kepler's second law is that the areas swept by the radius vector from the sun to the planet, are equal in equal time, throughout its orbit. This results in the planet moving fastest when near the sun, and slowest when far from the sun. For example, each day the motion of the planet sweeps out the same amount of area.

His third law relates the time it takes the planet to make one revolution around the sun, P, to its average distance from the sun, R. He found that the length of the planet's year squared is proportional to the distance of the planet from the sun cubed, or $P^2 = kR^3$.

The Gyroverse
The Hidden Structure of the Universe

ISAAC NEWTON

Isaac Newton, born 1642, the same year Galileo died, was an English scientist and mathematician. He was one of the most brilliant scientists the world has ever known, and its first physicist. He explained how gravity controls the motion of planets and stars, and keeps us bound to Earth.

Sir Isaac Newton was the father of modern physics. His treatise in the book <u>Principia,</u> 1687, included his three laws of motion, which for most real life problems still suffice today. He and Gottfried Wilhelm Leibniz simultaneously invented Calculus, the most important tool of modern mathematics. Calculus was needed to solve most problems involving the laws of motion. In addition, he discovered the universal law of gravitation, a formula for the gravitational attraction between two massive bodies. Finally, he solved many longstanding problems of motion within the solar system, such as, the causes of the tides, the precession of Earth's axis, and the effect of the sun's gravity on the motion of the moon.

The formula for universal law of gravitation is:

$$F = G(m_1 m_2 / R^2)$$

Where F is the force of attraction between the two masses m_1 and m_2, and R is the distance between their center points. G is the universal gravitational constant. It is universal because it is the same for all bodies in the universe.

Newton also corrected Kepler's third law. Newton realized that Earth does not revolve around the sun, but that the sun and Earth revolve around a common point between them. However, because the sun is so much more massive, the common point is very close to the sun's center, making Newton's correction small.

An important observation is that time does not enter into the gravitation formula. By implication, it appears to be acting instantaneously along the line adjoining both objects. Light takes more than eight minutes to make the trip from the sun to Earth. If eight minutes were added to the force delay, it would cause large errors in the orbits.

Donald Wortzman

STANDARD COSMOLOGY MODEL

GENERAL RELATIVITY

Albert Einstein published his theory of general relativity in 1916, eleven years after publishing his special theory of relativity. This theory was based on the postulate that gravity and acceleration were indistinguishable. This led to the theory that gravity was not a force as Newton described, but merely a manifestation of free motion in a curved space-time. Einstein's theory was made more profound because, unlike many other theories, no one else was thinking along similar lines. The solution of these equations for the universe included a constant that by default was set to zero. With the constant set to zero, the solution required that the universe be expanding or contracting. To be consistent with the paradigm of the day that the universe was static, neither contracting nor expanding, he introduced a non-zero cosmological constant to counterbalance the contracting force of gravity. Edwin Hubble later determined that the distant galaxies were receding, contradicting the static universe. Einstein subsequently regretted its inclusion and removed it, calling it "the biggest blunder in my life." However, the idea of the constant has never really vanished. In fact, one contemporary theory motivated by the accelerated expansion of the universe reintroduced the constant.

Einstein proposed the cosmological principle that declared that the universe is homogeneous and isotropic, looking the same everywhere and in all directions. Over a smaller scale, the universe is non-uniform, since between the super-clusters of galaxies there are wide voids without apparent matter. On a larger scale, it does appear that the principle holds. This is the accepted theory of most modern cosmologists.

HUBBLE CONSTANT

Hubble was first to measure that the universe was expanding uniformly, confirming the Friedmann-Lamaitre solution to general relativity equations. In other words, all matter in the universe is receding from all other matter at a rate proportional to their distance apart from each other. The equations for a distant galaxy are:

$$v = Hd$$
or
$$d = v/H$$

Where H is the Hubble constant, d is the distance from Earth to the galaxy, and v is the velocity that the galaxy is moving away from Earth.

Stars have two separate motion components. The first is a general expansion that is uniform; the other is the individual motion of each star. For stars that are very distant from Earth, billions of light-years away, the expansion component dominates. If the expansion (recession) velocity is measured knowing the Hubble constant, one can estimate the star's distance.

Using the Hubble constant, it is possible to extrapolate backward to estimate the time of origin of the universe. As the Hubble constant was refined, time estimates of the big bang improved. The belief that the universe was expanding faster at first, then slowed by gravity, lead to a lower age estimate. Recent measurements that the expansion is accelerating, increased the estimate. Consequently, recent estimates of age of the universe lie between 13.5 and 14.5 billion years old. The current most widely presented estimate is 13.7 billion years.

HUBBLE CONSTANT UPDATE

In the most recent determinations of the Hubble Constant two modifications predominated. First, the accuracy of measuring distance and velocity of galaxies has been expanded and improved. Secondly, General Relativity with accelerated universe expansion has been factored in. Consequently, the most current calculations by one group pegs the Hubble Constant, H at 73.2 kilometers per second per megaparsec, and the age of the universe, its reciprocal, T at 14.6 billion light years. Another group that starts with the

CMB era gets H to be 67.4 k/s/m and T to be 13.5 bly. Both groups are very sure of themselves. However, since the Gyroverse does not support GR or accelerated expansion the traditional calculation with current arithmetic is applied getting H to be 70 k/s/p and T at 14.0 bly. While most references refrain from calling it the Hubble Constant, because they accept the accelerated expansion, the Gyroverse still goes with Hubble Constant.

BIG BANG THEORY

In 1922, Alexander Friedmann, a Russian scientist, found a solution to general relativity, different from Einstein's solution. It called for a universe that expanded originally from a point, a finite time ago. Georges Abbe Lamaitre (Belgian) independently rediscovered the solutions, previously found by Friedmann, a few years later, popularizing the theory. The "big bang" model was further justified by George Gamow in 1946, a Russian-born U.S. nuclear physicist, who explained the proportion of atomic elements observed today. The term was coined by Fred Hoyle, a steady state proponent, who used it derisively to describe the creation event. The name stuck because it captured everyone's imagination. Friedmann communicated this work to Einstein. At first, Einstein was skeptical of this solution to the general relativity equations, but after more communications with Friedmann and some associates, he realized that it was a valid solution.

According to the big bang theory, the universe could either, expand forever, if its gravitational attraction was not sufficiently strong to reverse the outward motion of galaxies, or it could reach a maximum expansion and then start collapsing back to the single point, whence it came. To decide which, depends on the density of matter in the universe as compared with the rate of expansion. Current density measurements favor the "expansion forever hypothesis."

Another argument in favor of the big bang theory is that it predicts the relative abundance of elements in the universe that is consistent with observation. It predicts that the universe contains mostly hydrogen and helium in the proportion 75% to 25% (by weight) with trace quantities of everything else. This distribution is confirmed by observation of the characteristic frequency spectrum of starlight.

STELLAR DISTANCE

All the stars in the night sky seem to be the same distance away. This is why for 2000 years, all the stars were believed to be on a huge celestial crystal sphere that had a daily westward motion. This represents the rising and setting of the stars, caused by Earth's eastward rotation. Now we know that the stars are all far and different distances from Earth.

The two most common units to measure the distance to stars are light-years, in popular literature, and parsecs, among professional astronomers. A light-year is the distance that light travels in one year. The closest star Proxima Centauri is about 4.2 light-years away. The furthest observed are about 13.5 billion light-years away. A parsec, based on a triangular measurement is equal to 3.26 light-years.

Three techniques for estimating the distance to celestial bodies are used. These are parallax shift, a form of triangulation; standard candles, stars whose luminosity can be estimated; and red shift of light from celestial bodies. Each is best suited for different distance ranges.

Parallax Shift

Stellar parallax or parallax shift is the earliest common method for estimating star distances from Earth. In surveying, a similar method is referred to as triangulation. In this technique, a star is viewed at a six-month interval, when Earth is furthest to one side of the star, and six months later when it is furthest to the other side of the star. Depending on the star's position in the sky, different six-month intervals will maximize the separation. The relative location to another star that is much further away is compared, noting the apparent angle of separation at both extremes. From the difference of the two angle measurements, and the distance between the two Earth positions at the six-month intervals, the star's distance from Earth can be estimated. This method works best for stars that are close to Earth, less than 100 light-years away. Otherwise, the angles are too small to be accurately measured. This is made clear by referring to the following diagram.

Figure 10.1 – Parallax Shift

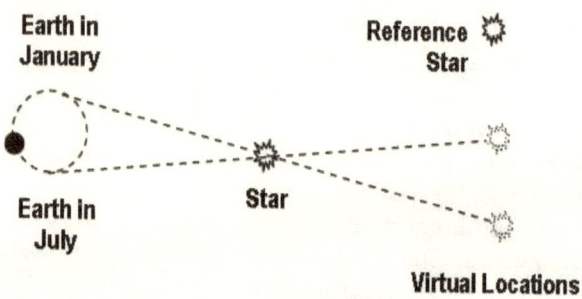

Figure_10.1 - This figure is illustrative of the parallax technique. On the left Earth is shown orbiting the sun (not shown). The star in the center is viewed at a six-month interval. By comparing its apparent virtual location at these times with respect to the reference star the angle difference between both views can be measured. This is sufficient for calculating the distance the star is from Earth. Note that this illustration is not drawn to scale. In particular, the reference star must be orders of magnitude further from Earth than the star whose distance is to be measured.

Standard Candle-Cepheid Variables

A standard candle is a bright celestial object that has a predictable luminosity. In 1912, Henrietta Leavitt, studying the small Magellanic Cloud, one of the closest galaxies to the Milky Way, published her work on Cepheid variables, used as standard candles. These stars having used up their stock of hydrogen fuel pulsate between dim and bright repeatedly. Leavitt discovered that the average luminosity of these stars is related to their periods and is 10 to 20 thousand times brighter than the sun. Therefore, if a Cepheid variable's period is measured, its absolute luminosity can be inferred. If the apparent brightness of the Cepheid variable is measured on Earth, using the formula that the star's brightness is proportional to its luminosity divided by the square of its distance, then the distance can be estimated. In equation form:

$$D2 = L/kB$$
$$\text{or}$$
$$k = L/BD2$$

Where L is the calculated Luminosity for that period Cepheid, B is the measured brightness of the Cepheid, D is the distance to the star, and k is a constant of proportionality.

The constant of proportionality can be determined by applying this formula to a Cepheid whose distance can be estimated by other means. For example, the "original Cepheid variable," Delta Cepheid, is close enough that we have a parallax measurement for its distance from Earth. The Cepheid standard candle has been invaluable for estimating relatively nearby galaxies, to about 20 million light-years away from Earth.

Standard Candle-Supernova Type 1a

Another example of a standard candle is a Type 1a supernova. These supernovas are certain white dwarfs near the end of their life that go through a swift huge nuclear reaction that produces immense predictable peak luminosity. These white dwarfs accumulate mass from a companion star. When its mass reaches about 1.4 times the mass of the sun, known as the Chandrasekhar limit, named for Subrahmanyan Chandrasekhar, the Nobel Laureate who derived it, they undergo an immense nuclear reaction, outshining the galaxy there in. Their light output is so bright that they can be seen almost back to the origin of the universe, 500 times further than Cepheid variables. If the peak light output, about four billion suns of a supernova in a distant galaxy is measured, by use of the inverse square law, the distance of its parent galaxy can be deduced. This mechanism may be compared with a hydrogen bomb the size of Earth, having the mass of the Sun. These have been used in estimating distances to galaxies to about 10 billion light-years away, and played a major role in the studies that determined that the universe was in an accelerated expansion. This was done by comparing distance estimate bases on its intensity as well as its red shift.

Doppler & Expansion Shift

Atoms of the same element have a unique pattern and emit light that consists of a definite set of colors. Each color corresponds to a different wavelength of light. Because an element is associated with a unique pattern, recognizing the types of atoms in distant galaxies by measuring the wavelengths of the light they emit is possible by matching the measured wavelength values with the known element pattern.

When light is emitted from an object that is moving away from an observer, the speed of light to the observer does not change, but remains c. In contrast, a Doppler shift causes the frequency of the light emitted to decrease, and its wavelength to increase. The increase in wavelength is proportional to the speed of recession between the object and the observer. This increase in wavelength is called red shift, because red light is a longer wavelength photon. The recession velocity is roughly proportional to the percent red shift change times the speed of light, which is:

$$v = c\,(\lambda_o - \lambda_e)/\lambda_e$$

Where λ_e is the emitted wavelength, λ_o is the received wavelength, v is the recession velocity, and c is the speed of light.

Remembering that the distance is equal to the velocity divided by the Hubble constant, the equation is approximately:

$$d = v/H = c(\lambda_o - \lambda_e)/\lambda_e H$$

When Hubble looked at the wavelength pattern of the light emitted from distant galaxies, he could recognize the elements because all the wavelengths were longer proportionally. He could then calculate the red shift and get estimates for the galaxy distance from Earth. For distant stars, this is the preferred way to calculate its distance from Earth. For closer stars, the unique motion separate from their general recession velocity has a big impact on the recession calculation. Therefore, the distance estimate would have larger errors.

While red shift is referred to as Doppler shift, strictly speaking the frequency shift is considered to be due to the general expansion of the universe and not the Doppler effect. The Doppler shift only depends on the relative speed of the source when the light is emitted and the receiver when the light signal is captured. It does not matter what occurs in between. On the other hand, expansion shift only depends on the relative size ratio of the universe between the time the light was emitted and when it was received, and not the initial and final relative speeds.

Since all matter is moving away from all other matter proportional to their distance apart, objects far beyond each other's horizon must be moving away from each other much faster than the speed of light. Consequently, it is assumed that stretching of space, not conventional motion, is causing this red shift of distant galaxies. Overall, the expansion of the universe is due to the expansion shift, and the local variation is taken to be the Doppler shift. Both paradigms lead to the same result, if the expansion rate is constant, very close to what has occurred for most of the universe's existence.

INFLATION MODEL

The uniform expansion of the universe highlights two problems: the flatness problem and the horizon problem.

Flatness Problem

The flatness problem comes about because calculations show that for the universe to have evolved as it did, the big bang expansion force had to be balanced to an unlikely high precision with gravitational attraction. A measure used to determine the flatness is defined by the parameter omega (Ω). If Ω equals exactly one, then the universe would be perfectly balanced, its expansion slowing to zero, but never reaching it. If it is greater than one, then the universe will eventually contract. On the other hand, a number less than one means the universe will expand at its escape speed forever. Best estimates today are that Ω is between 0.1 and 2.0. Working backward, this implies that Ω had to be almost exactly one. If the expansion parameter, Ω, initially were any larger, the universe would have collapsed long ago, or if it were smaller, matter would not have clumped together to form the stars and galaxies. The initial expansion had to be balanced with the gravitational attraction perfectly to within one part in a trillion for the universe to be as flat as it is today--a very unlikely possibility.

Horizon Problem

The horizon problem became known when the background radiation coming from space in all directions was measured and found almost identically the same. This cosmic microwave background radiation is a remnant of the early years of the universe, when it was 300 thousand years old; the time after which radiation could travel uninterrupted in the universe. For the radiation in all directions to be of uniform intensity now, it shows that matter at the outer reaches of the universe in all directions must have been homogeneous in the distant past. This is highly unlikely, with the universe's expansion rate predicted by the general relativity Friedmann solution. Regions where the radiation came from would never have been in causal contact with each other. For example, the universe observed at distances of 13.5 billion light-years in each direction, indicated that light would take at least 27 billion years to travel from one end to the other. These places could not have been in contact since the universe is thought to be between 13 and 15 billion years old. Yet, the radiation uniformity in all directions inferred that they must have been in contact. The dominant big bang theory, being in conflict with these results, made the universe expansion hypothesis problematic.

In 1980, Alan Guth, Professor of Physics at the Massachusetts Institute of Technology, proposed another theory, called the inflation model. Unexpectedly, and to quantum physicists' delight, his solution relies on some well-known quantum physics' principles, quantum fluctuation. This allowed instant creation of energy bubbles, composed of a minute amount of particle antiparticle pairs. These would normally have disintegrated immediately was it not for an immense gravitational reversal causing an outward expansion force that initiated the universe inflation period. During this inflation, the universe doubled in size more than 100 times in less than 10^{-33} seconds. This inflation caused all the matter in the universe, which has positive energy, to be created. The energy of matter was exactly counterbalanced by the negative gravitational energy. Accordingly, the universe has a net of zero energy. Edward Tryon, a professor at Hunter College of the City University of New York, proposed in 1973 that the universe could have started from nothing by a quantum fluctuation. He described it as "simply one of those things that happens from time to time." Guth's concept, built on top of Tyron's idea, instantly received much attention. However, it suffered from an inability to gracefully stop the inflation process. Andrei Linde, a Russian cosmologist, and **Paul Steinhardt, and Andreas Albrecht,** working together at the University of Pennsylvania, proposed an improved formulation that got around the difficulties with Guth's model. The development of this theory is chronicled in a book written by Alan Guth, The Inflationary Universe. In 2002, Guth, Linde, and Steinhardt were co-recipients of the prestigious Dirac Medal for their work on inflation theory.

Although Guth started this work to solve the magnetic monopole problem, he recognized that General Relativity has an implied relationship between Ω and the geometry of the universe, where the distribution of matter in the universe determines its geometry. A uniform distribution of matter produces a flat or Euclidean universe. Non-uniform distributions can cause a concave or a convex universe.

In a concave universe, space closes in on itself. Straight lines in any direction will return to the starting point. The two-dimensional analogy is Earth, where every direction brings one back to the starting point. Because of the correlation between the distribution of matter and the geometry of the universe, the concave closed condition would be recognized by the abundance of matter increasing in density as the distances get greater, making Ω greater than one. A convex, forever expanding universe, is the exact opposite, and would be recognized by the density of matter reducing as the distance away gets larger, making Ω smaller than one. A flat universe is the critical transition between concave and convex, with Ω exactly one. Guth recognized that inflation would make the geometry flat, which would also drive Ω closer and closer to unity initially, enabling it to still be relatively close to unity in the present epoch. These three conditions are characterized by the following figure.

Figure 10.2 – Three Universe Expansion Solutions

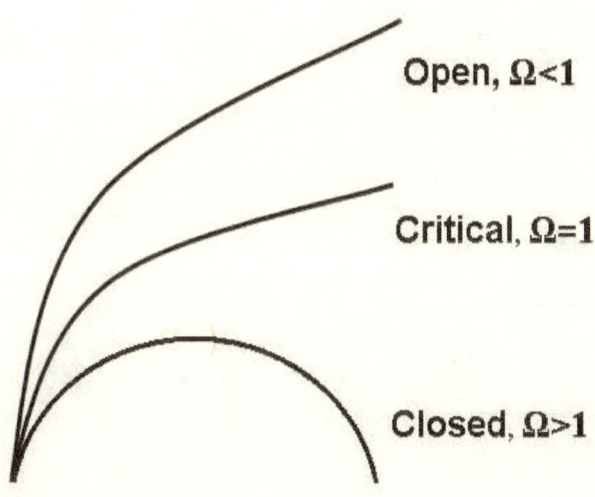

Figure_10.2 - This figure illustrates the three possible expansion scenarios of the universe. The lower curve depicts a universe that initially expands, but ultimately contracts due to gravitational attraction of all matter. The top curve portrays a universe slowed by gravity, but not sufficiently to reverse the expansion. It approaches a maximum expansion speed that continues forever. The middle curve shows the situation where the outward expansion is perfectly balanced between the two extremes. Here, the expansion speed approaches zero, but does not reach it. In reality, this critical case cannot exist because the amount and distribution of matter must be perfect. If the gravitational force is slightly too large, the universe will ultimately collapse. If it is slightly too small, it will default to the open case.

In the first fraction of a second, the universe grew much larger than the observable universe is today. A two-dimensional analogy would be confetti on a balloon moving away from each other as the balloon is inflated. The inflation model could resolve the flatness and horizon problems, because the total expanse of our observable universe originated from a very small portion of the original big bang that was exceedingly flat. As already stated, according to Guth, during inflation matter was created from the gravitational energy, which is negative. This inflation caused the universe to grow exponentially, such that in 10^{-33} seconds after the big bang it doubled in size more than 100 times.

Getting back to the balloon analogy, our entire viewable universe would encompass just a small surface section of the balloon. This small section would have been in intimate contact just after the big bang, so that it would not be surprising that they exhibit identical characteristics. Inflation theory allows the universe to grow with distances between matter uniformly increasing, without matter moving in the conventional sense. Actually, this inflation model introduces a third theory to supplement quantum and relativity theories. While the model is widely accepted, it is still being revised.

Accelerated Universe Expansion

Recently, it has been learned that not only is the universe expanding, but also its expansion is accelerating. This is inconsistent with expectations, since it was assumed that gravity would work to slow the expansion, even if it were not great enough to halt it. Studies that discovered the accelerated expansion were originally initiated to measure the gravitational slowing of the expansion. Two teams in 1998 almost simultaneously measured this acceleration in expansion speed. Saul Perlmutter of Lawrence Berkeley National Laboratory's Physics Division headed one group. The second group was lead by Robert Kirshner of Harvard University Center of Astrophysics, where he was a professor. Kirshner wrote The Extravagant Universe, an exciting book that recorded this competitive saga.

This conclusion of accelerated expansion is arrived at by combining two different distance measurements on supernova of Type 1a. If the peak light output of a supernova in a distant galaxy is measured, by use of the inverse square law, the distance of its parent galaxy can be deduced. Alternately, the parent galaxy distance can be indirectly measured using the more standard red shift. When this dual measurement procedure is applied to this Type 1a supernova, the luminosity distance measurement infers that the galaxy is further from Earth than the red shift measurement suggests. For distant galaxies, the distance difference becomes progressively greater. Since the red shift measurement inferred the receding velocity of the galaxy to be less than predicted, it implied that the galaxy's velocity was less in past times and must be accelerating now. This was an unexpected result because it was always assumed that gravity was slowing the universe's expansion. Now, to reconcile this new information with the previous assumptions of the evolution of the universe, a cosmological constant had to be added again to general relativity. This time it is needed to explain the accelerated expansion of the universe, not just keeping it static. Unexpectedly, it implies that empty space is not empty but embodies considerable energy to provide an expansion force. Best recent estimates are that the matter in the universe account for a contribution to Ω of about .3, approximately 80% of which was thought to be dark matter. This was not as close to an ideal $\Omega=1$ flatness that cosmologists expected. Now the dark energy (mass equivalent) contribution to Ω is thought to be roughly 7. Together, they explain flatness without the need to discover more dark matter. Nevertheless, the cosmological constant is troubling because it not a constant of nature but is an ad hoc solution to the accelerated expansion problem. In addition, the current understanding of how vacuum energy and gravity relate, predicts 10^{120} times more dark energy than observed.

Donald Wortzman

PART III – THOUGHT EXPERIMENTS

Donald Wortzman

CHAPTER 11 - TRIPLETS PARADOX

INTRODUCTION

Einstein introduced a famous thought-experiment, known as the Twin Paradox. In this scenario, one twin traveled away and then back to Earth at a large fraction of the speed of light while the other remained on Earth. Einstein explained that the twin that traveled got back to Earth much younger than his twin that stayed put. This experiment is considered a paradox because Special Relativity gives the impression that either one can be considered at rest, with the other moving. The difference of course is that the traveling twin experiences several accelerations and decelerations that differentiate the two circumstances. Faulting this explanation is difficult, because the twins are far apart most of the time, and neither can effectively dispute the other's experience. By adding a third brother to this paradox, the error in this rationalization becomes apparent.

Two of the three triplets, Tinker and Evers live on Earth. The third triplet, Chance, sells real estate on Soil an Earth-like planet orbiting Alpha Centauri, 4 light-years away. Timepieces have been synchronized between Soil and Earth. Tinker plans to visit his brother Chance, on Soil, leaving on the first of January.

SCENARIO

At 12 a.m., January 1, Evers on Earth sends a space-mail message to Chance on Soil that his brother Tinker is just leaving to visit him by spacecraft. In eight hours, spacecraft time, Tinker achieves final cruising speed of .8 times the speed of light. He sends a space-mail message to his brother Chance stating the final cruising speed of the spacecraft. It is routed by way of his other brother Evers who is on Earth and has a more powerful transmitter. Evers gets the message later that first day, timestamps it, and forwards it to Chance. Both messages are received by Chance about one year before Tinker's expected arrival, which according to Chance will be December 31, 5 years after Tinker's departure.

A few days before Tinker's expected arrival at Soil, he sends a space-mail message to Chance requesting authorization to dock on Soil. Chance, 24 hours before the expected docking time, sends back a space-mail message authorizing Tinker to decelerate and dock, with which he complies.

Conclusion

According to all three triplets, except for the first and last day, the spacecraft had a constant speed toward Soil at .8 times the speed of light. Because of the symmetry of that portion of the trip, the trip took the same 5 years for all three triplets, plus or minus the uncertainty time contribution of the first and last day, where there was acceleration.

One could similarly argue that the trip took the same time for all three triplets, but since it took the spacecraft 3 years to make the trip, the trip took the same 3 years for all the triplets. The point is that not all methods of calculation lead to the same answer.

Traditional Solution

Traditionally the elapsed time of the trip would be 5 years for Chance and Evers, who both are on the same inertial frame and have experienced no acceleration or deceleration. Since Tinker did first accelerate when he left Earth and decelerated to arrive at Soil, he would have experienced the time dilation of the amount:

$$\gamma = (1-v^2/c^2)^{-1/2} = (1-[.8c]^2/c^2)^{-1/2} = (1-.64)^{-1/2} = .6$$

The time of travel for Tinker would be:

$$\gamma(5 \text{ years}) = .6(5 \text{ years}) = 3 \text{ years}$$

If the theory is correct, all reasonable applications of the theory should give the same answer. The problem is that in special relativity equality is not transitive. (This principle states that if A = B and B = C, then A = C.) Thus, it makes a difference how one chooses to solve this problem.

COMMENTS

Special relativity predicts that according to Tinker's (spacecraft) timepiece, the trip took 3 years and he traveled 2.4 light-years. According to Evers' (Earth) and Chance's (Soil), the trip took 5 years for a distance of 4 light-years. Oddly, Evers and Chance do not agree on when Tinker reached final speed. The special relativity equation for translating time events between inertial frames in relative motion is $t=\gamma[t'+vx'/c^2]$, where the primed variables are that of the spacecraft traveling at .8c, and the unprimed are that of Earth's inertial frame. For example, the time that the spacecraft reaches .8c as viewed by Earth is the first day ($t'=0$, $\gamma=1.66$, $V=.8c$, $x'=0$). On Soil, it is 3.2 years ($t'=0$, $\gamma=1.66$, $V=.8c$, $x'=2.4$).

This scenario was structured such that the key events happened on Earth's and Soil's common inertial frame. The importance of this is that relativity should not be playing a role in the timing of the events. All three observers would agree to the following:

1. Earth time of departure - First day of trip.

2. Earth time when Earth received the message that the spacecraft reached final speed - First day of trip.

3. Soil time when the message was sent to the spacecraft to dock on Soil - Final day of trip.

4. Actual docking on Soil - Final day of trip.

Elementary logic would presume that since Earth had a message the first day that the spacecraft reached cruising speed, the spacecraft must have reached cruising speed before the message was received. According to special relativity, Soil places the event 3.2 years after Earth received the message that it happened.

In both special relativity and the gyroverse theory, clock time would slow the same amount, so that is not the issue here. The issue is the explanations surrounding the slowing of clock time. In the gyroverse theory, as the spacecraft builds up speed, the timepiece in the spacecraft slows in relation to how fast it is moving. All observers would agree on all details, such as departure time, arrival time, clock times of all events, and aging of the players, after adjusting for different clock rates. In special relativity, the occupants, by theory, must age exactly in relation to the slowing of the timepiece, since real time is actually slowing. In the gyroverse theory, whereas clock time slows, and atomic processes slow proportionally, real time is not slowing. Thus, it is not certain whether aging or other biological processes will exactly match the slowing of clock time, although they will surely slow somewhat. What makes this scenario different is that all messages are routed by way of the triplet that is on the stationary inertial frame, so that everything can be compared time wise, simply from the same frame.

GYROVERSE EXPLANATION

A traveler moving at .8c between Earth and Soil 4 light-years apart would take 5 years for the trip from the perspective of Earth and Soil, but 3 years from the perspective of Chance. By traveling 4 light-years away, but clocking only 3 years, he has traveled faster than the speed of light, from that perspective. Nevertheless, at .8c, and a gamma of 1/.6, Earth and Soil appear 40% closer, because of their difference in yaw with respect to the spacecraft. Chance's instruments would still show the speed to be .8c, but the distance, appearing 40% shorter, makes the trip appear shorter. His clock running slower gets him there in 3 clock years, as far as he is concerned. Albeit, his clock is running 40% slower as compared with clocks on Earth and Soil, resulting in the trip really taking the same 5 years, although he probably aged less. In addition, the distance looked shorter, but it really was not.

CHAPTER 12 – INSTANT MESSAGING

INTRODUCTION

Described is a communication mechanism that is faster than the speed of light, almost instantaneous. It has been shown that entangled particles can correlate with each other much faster than the speed of light. Even though they are correlated, transmitting messages using this correlation is not possible, according to orthodox theory. The reason is that all entanglement only seems possible while the particles are in random state. The mere attempt to constrict the particles to any non-random subset seems to destroy their entangled state. While no one has yet successfully succeeded in demonstrating faster-than-light information transmission using this principle, it is not clear that it cannot be done.

A related action-at-a-distance phenomenon to entanglement is the ability of a single atomic particle to be dispersed over large distances. An example is the double-slit experiment, where a single photon appears to pass simultaneously through two slits. Again, it seems plausible that this peculiarity of nature could lend itself to faster-than-light information transfer. In the book's first and second editions, the methods for achieving faster-than-light transmission had flaws. In this third edition, the problems encountered were fixed using ideas borrowed from Wheeler's delayed choice scheme and Gisin's 7 mile entanglement experiment.

DESCRIPTION OF EXPERIMENT

The schematic to achieve this almost instant messaging is shown in Figure 12.1. Monochromatic light originates at the photon source. Although it is not necessary, assumed for this initial discussion is that the light-beam is of low enough intensity so that the photons are emitted individually.

The first 1X2 splitter is symmetrical so that half the photon beam goes through the right fiber optic cable and the other half goes to the left. The beam going left meets another juncture, a phase splitter, where a quarter of the light goes each way. The phase splitter produces two simultaneous beams, 180 degrees out of phase with each other. Its operation is described in detail later with the help of Figure 12.2. On the right, the beam goes through an electro-optic delay switch. The voltage on the electro-optic delay switch modulates the fiber's index of refraction. Increasing the voltage increases the index, so that at some voltage, the delay increase reaches a half wavelength. At its lower level, the added delay is zero. Initially the input voltage to the electro-optic delay switch is at its lower level, reducing the delay to zero. With both legs equidistance, the photon beams arrive at the 2X1 combiner simultaneously out-of-phase with each other, destructively interfering, causing all the photons to appear at the right output detector. The output voltage of the left detector is at its low voltage, and the right detector is at its high voltage, setting the output latch. Consequently, the inverted latch output is at its lower level. Summarizing, when the input voltage is at its lower level, the inverted latch output is at its lower level.

If the electro-optic delay switch input is subsequently set to its upper level, the delay through that juncture is delayed half a wavelength. Hence, the light wave meets at the 2X1 combiner at different times and constructively interferes, allowing half the light-beam to pass to the left detector, generating half voltage output at the left photon detector output. Since half the photons enter the left detector, only half arrive at the right detector, lowering the output voltage at that detector. This smaller output voltage flips the latch, setting its output to its upper level. In summary, when the optical delay switch's input voltage is at its upper level, the output voltage is also at its upper level.

Assume that both detectors are equidistant from the photon source, but tens of meters from each other. Also, assume that the electro-optic delay switch is less than a meter from the left detector. Then, neglecting the switching, and other circuit delays, once the switch is thrown from constructive to destructive interference, all photons upstream from the switch will only be detected at the right detector, doubling its output. When the switch is thrown the other way, the right detector goes back to the original output. A photon that is centimeters from the switch will then appear or disappear at the right detector, tens of meters away, in the time it takes light to travel about a meter. The result is communication, much faster than light-speed. While for clarity, the description assumed that the photons were released one-at-a-time, the restriction was not used and is not necessary. Since this method alters only the probability of taking each path, many photons are needed to get the voltage change. Therefore, the intensity of light needs to be high enough, and the response of the detector fast enough, so that the time for a large number of photons to switch the detector is insignificantly short.

In addition, there are several variations of the setup that can be made.

1) Another delay switch with phase splitter and combiner could be placed on the right side, and another latch placed on the left side, making the arrangement symmetrical, allowing messages to be sent in both directions.

2) While orthodox theory says that there is no theoretical limitation for the distance between detectors, Gisin has demonstrated entanglement to seven miles. It follows that this alternate action-at-a-distance phenomenon would most likely also work to that same distance.

3) The output of the latch can be fed to the input of another detector so-as-to repeat the signal, sending it double the distance. In fact, any number of repeaters can be used to send the signal any distance.

The orthodox explanation, Copenhagen interpretation according to Wheeler, is that the photons go back in time and revert to going down the right side only when they find the left side blocked by destructive interference. In the gyroverse interpretation, each photon is simultaneously taking both paths in sister-locations. While these sister-locations may be 7 miles apart in three-dimensional space, they are actually 10^{-30} cm apart in the full twelve-dimensional space, close enough that each photon can simultaneously span both paths. When it destructively interferes on the left side, it concentrates on the right side where there is no interference. Nothing ever goes backwards in time.

Figure 12.1 - Instant Messaging – Faster Than Lightspeed Transmission Configuration

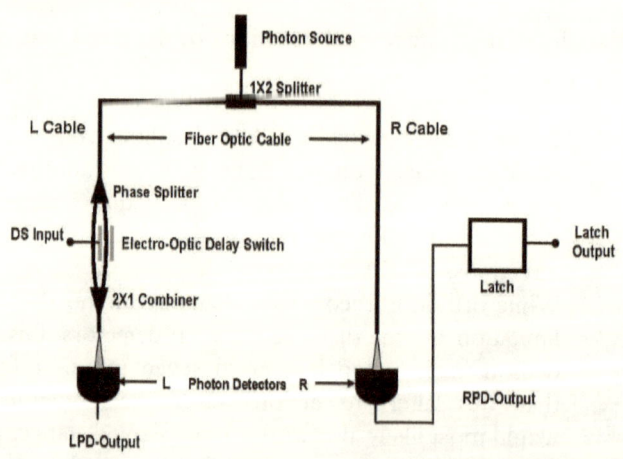

Figure_12.1 - A photon source feeds a splitter, which allows a single photon to split between flowing down the left input side and the right output side. Each detector is equidistant from the photon source, at least tens of meters away, but the electro-optic delay switch is less than a meter from the left detector. On the left, the photon enters a phase splitter, which produces an in-phase and inverted phase signal. With the delay switch at zero delay, both signals arrive together at the combiner with opposite phase, canceling each other out. With the delay switch set to half a wavelength, no cancellation results and the signal passes unabated through the combiner. In this mode, approximately, fifty percent of the photons register at each detector, which are equidistant from the photon source. In the former mode, all the photons appear at the right output-side detector, and no signal appears at the left input-side detector.

Varying the probability of the outcome by increasing the input voltage can be delayed to the point where the photons have already traveled a long distance. Consequently, in the time it would ordinarily take a photon to travel the short distance from the switch to the left-side detector, 50% of the photons can be diverted from the left input-side detector to the right output-side detector, a long way from the delay switch. However, it takes many photons to ensure that the change has taken place.

Figure 2 is a schematic of the phase splitter. The photons enter the phase splitter from the top and strike the half-silvered mirror. Half the light-beam passes through the mirror and through the light funnel which concentrates the light-beam into a fiber cable. The other half of the beam bounces off the half-silvered mirror with a phase reversal, then passes through the glass slab, and finally concentrated by the light funnel into fiber cable. Since both paths are of equal length, passing through an equivalent amount of material, the beams enter fiber-optic cable simultaneously, but 180 degrees out of phase.

Figure 12.2 - Phase Splitter

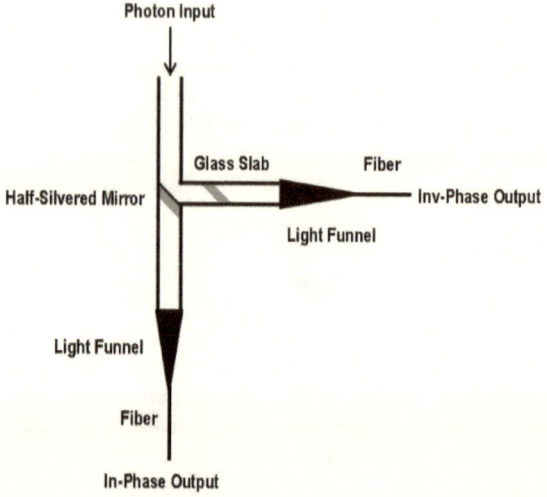

Figure_12.2 - A photon source feeds the phase splitter, which allows a single photon to split between flowing down the left and off to the right, bouncing off a half-silvered mirror, which causes phase reversal. Both signals arrive at their destination at the same time but 180 degrees out of phase.

CONCLUSION

The argument that restricts quantum entanglement from being used for information exchange is that each set alone is random. Only by comparing both sets does the information being transmitted show through.

A similar argument applies to a single photon taking multiple paths. Since it cannot be predicted which particular path a photon will take, using this characteristic to transmit information is also not possible.

While it is not possible to predict which path a particular photon will take, altering the probability of taking each path is possible. The thought-experiments in the book's first and second editions were attempts to use this probability alteration on entangled photons to transmit messages, but both had flaws. This third attempt corrects those previous flaws.

When a photon stream is split between two paths, it's not that each photon takes one or the other paths, but each photon takes partially both paths simultaneously, but end up in just one place. Some photons terminate prematurely. Where the others terminate is a function of what they meet along the way. By placing a variable barrier in one path, the final destination of the photons can be altered. However, it is only varied statistically, so it is not possible to know what path any particular photon takes. This arrangement is unique because it relies on this statistical characteristic, where previous attempts for transmitting information try to predict the paths of individual photons.

BIBLIOGRAPHY

Abanes, Richard – **The Truth Behind the DaVinci Code:** A Challenging Response to the Bestselling Novel - Harvest House Publishers / 2004

Abbott, Edwin A. – **Flatland: A Romance of Many Dimensions** – Signet Classic / 1984

Aczel, Amir D – **Entanglement:** Unlikely Story of How Scientists Mathematicians, and Philosophers Proved Einstein's Spookiest Theory – Plume Book / Oct 2003

Aczel, Amir D - **God's Equation:** Einstein, Relativity, and the Expanding Universe - Dell Books / December 2000

Ananthaswamy, Anan – **The Edge of Physics:** A Journey to the Earth's Extremes to Unlock the Secrets of the Universe – Harcourt Publishing / 2010

Armstrong, Karen - **The Battle For God:** A History of Fundamentalism – A Ballentine Book / 2000

Armstrong, Karen - **The Bible:** A Biography Atlantic Monthly Press / 2007

Baggott, Jim - **The Meaning of Quantum Theory:** A Guide for Students of Chemistry and Physics - Oxford University Press, Incorporated / May 1992

Baigent, Michael, & Leigh, Richard, & Lincoln, Henry – Holy Blood, Holy Grail – Delacorte Press / 2004

Baigent, Michael & Leigh, Richard – The Dead Sea Scrolls Deception - Simon & Schuster / 1992

Bainbridge, David – The X in Sex: How the X Chromosome Controls Our Lives – Harvard University Press / 2003

Banes, Richard A. – The Truth About The DaVinci Code – Harvest House Publishers / 2004

Barash, David P. – The Survival Game: How Game Theory Explains the Biology of Cooperation and Competition – Times Book / 2003

Bardi, Jason Socrates – The Calculus Wars: Newton, Leibniz, & The Greatest Mathematical Clash of All Times – Thunders' Mouth Press / 2006

Barrow, John D. - **The Book of Nothing:** Vacuum, Voids, and the Latest Ideas About The Origin of the Universe - Pantheon Books / 2000

Barrow, John D. – The Constants of Nature: From Alpha to Omega: The Numbers That Encode The Deepest Secrets of the Universe - Pantheon Books / 2002

Barry, John M – The Great Influenza: The Story of the Deadliest Pandemic in History – Penguin Books / 2004

Barry, Steve – The Third Secret – Ballantine Books / 2005

Barry, Steve – The Templar Legacy – Ballentine Books / 2006

Bergmann, Peter Gabriel - **The Riddle of Gravitation** - Dover Publications / January 1993

Behe, Michael J – **Darwins Black Box:** The Biochemical Challenge to Evolution – Simon and Schuster / 2006 & 1996

Behe, Michael J – **The Edge of Evolution:** The Search for the Limits of Darwinism – Free Press / 2007

Black, Edwin - **IBM and the Holocaust:** The Strategic Alliance between Nazi Germany and America's Most Powerful Corporation - Crown Publishing Group / 2002

Bock, Daniel L. – **Breaking the Davinci Code:** Answers to the Questions Everyone's Asking – Nelson Books / 2004

Bodanis, David – **$E=MC^2$:** The Biography of The Worlds Most Famous equation – Berkley Pub Group / 2001

Bohm, David - **Quantum Theory** - Dover Publications, Incorporated / May 1989

Bohm, David - **The Special Theory of Relativity** - Routledge / November 1996

Bolles, Blair – **Einstein Defiant:** Genius Versus Genius in the Quantum Revolution - National Academy Press / 2004

Boyne, Walter J. & Fields, Leslie Leyland & Smith, Fred - **The Two O'Clock War:** The 1973 Yom Kippur Conflict and the Airlift That Saved Israel - St. Martin's Press / August 2002

Brown, Dan – **Angels and Demons** – Atria Books / 2000

Brown, Dan – **DaVinci Code** – Doubleday / 2003

Brown, Guy C. - **The Energy of Life:** The Science of What Makes Our Minds and Bodies Work - The Free Press / April 2000

Brown, James Robert – **Who Rules In Science:** An Opinionated Guide to the Wars – Harvard University Press / 2001

Bruce Colin - **Schrödinger's Rabbit:** The Many Worlds of Quantum – Joseph Henry Press / 2004

Bryson, Bill – A Short History of Nearly Everything – Broadway Books / May 2003

Burnham, Terry & Phelan, Jay - Mean Genes: From Sex to Money to Food: Taming Our Primal Instincts - Perseus Book Group / August 2000

Burstein, Dan – Secrets of the Code: The Unauthorized Guide to the Mysteries Behind The DaVinci Code – CDS Books / 2004

Butz, Jeffrey J. – The Brothers of Jesus and the Lost Teachings of Christiananity – Inner Traditions / 2005

Caporale, Lynn Helena – Darwin in the Genome: Molecular Strategies in Biological Evolution – McGraw Hill / 2003

Carrol, Sean B. – The Making of the Fittest: DNA and the Ultimate Forensic Record of Evolution – W. W. Norton & Company / 2006

Carrol, Sean B. – The New Science of Evo Devo: Endless Forms Most Beautiful – WW Norton & Company / 2001

Casti, John L. - Paradigms Lost: Tackling the Unanswered Mysteries of Modern Science - Avon Books / November 1990

Chen, Fen - New Theory of Trisection: Solved the Most Difficult Math Problem for Centuries in the History of Mathematics - International School: Math and Science Institute / 1991

Clark, William R. & Grunstein, Michael - Are We Hardwired: The Role of Genes In Human Behavior - Oxford University Press / October 2000

Clegg, Brian – The God Effect: Quantum Entanglement, Science's Strongest Phenomena – St. Martin Press / 2006

Conant, Jennet – 109 East Palace: Robert Oppenheimer and the Secret City Los Alamos – Simon & Shuster / 2005

Cochran, Gregory & Harpending, Henry – The 10,000 Year Explosion: How Civilization Accelerated Human Evolution – Basic Books / 2009

Coulter, Ann H – Slander: Liberal Lies About The American Right – Random House / June 2002

Coulter, Ann H – Treason: Liberal Treachery From The Cold War to the War on Terrorism – Crown Forum / 2003

Coulter, Ann H – How to Talk to a Liberal (if You Must): The World According to Ann Coulter – Crown Publishing Group / 2004

Courant, Richard & Hilbert, David - Methods of Mathematical Physics, V1, V2 - Wiley, John & Sons, Incorporated / November 1990

Cox, Simon – Cracking The DaVinci Code: The unauthorized Guide to the Facts Behind Dan Brown's Best Selling Novel – Barnes and Noble Books / 2004

Cunningham, Philip A. & Thorpe, Kiki – Perspectives on the Passion of the Christ: Religious Thinkers and Writers Explore the Issues Raised by the Controversial Movie – Miramax Books / 2004

D'Aczel, Amir – Entanglement: The Greatest Mystery in Physics - Four Walls Eight Windows / October 2002

Danielson, Dennis Richard (Edited by) - **The Book Of The Cosmos:** Imagining The Universe From Heraclitus to Hawkings - Helix Book, Perseus Publishing - July 2000

Davies, Kevin - **Cracking The Genome:** Inside The Race to Unlock Human DNA - The Free Press / January 2001

Davies, Paul - **Are We Alone:** Philosophical Implications of the Discovery of Extraterrestrial Life - Basic Books / August 1996

Davies, Paul - **The Fifth Miracle:** The Search for the Origin and Meaning of Life - Simon & Schuster / 1999

Davies, Paul - **The Matter Myth** - Dramatic Discoveries: That Challenge Our Understanding of Physical Reality - Touchstone Books / 1992

Davies, Paul – **The Mind of God:** The Scientific Basis for a Rational World – Touchstone Books / 1992

Davies, P.C.W. & Brown, J (Editors) - **Superstrings - A Theory of Everything** (A Canto Book)- University of Cambridge Press / September 1992

Dawkins, Richard - **Climbing Mount Improbable** - WW Norton and Company / 1996

Dawkins, Richard – **The God Delusion** – Bantam Press / 2006

Dembski, William A. (Edited by)- **Mere Creation:** Science, Faith & Intelligent Design - Intervarsity Press / 1998

Denton, Michael J. - **Evolution: A Theory in Crisis:** New Development in Science are Challenging Orthodox Darwinism - Adler & Adler / March 1997

Denton, Michael J. - Natures Destiny: How The Laws of Biology Reveal Purpose in the Universe - The Free Press / 1998

Denton, Michael J. - Other Worlds: Space, Superspace, and the Quantum Universe - Pelican Books / 1980

Derbyshire, John - Prime Obsession: Bernhard Riemann and the Greatest Unsolved Problem in Mathematics – Joseph Henry Press / 2003

Deutscher, Guy – The Unfolding of Language: An Evolutionary Tour of the Greatest Inventions

Devlin, Keith - The Math Gene: How Mathematical Thinking Evolved and Why Numbers are Like Gosip - Basic Books (Perseus Book Group) / January 2000

DeWaal, Frans – Our Inner Ape: Why We Are Who We Are – Penguin Group / 2005

Dewdney, A.K. – Beyond Reason: 8 Great Problems That Reveal The Limits of Science – John Wiley & Sons / 2004

Dobbs, Lou – War on the Middle Class: How The Government, Big Business, and Special Interest Groups are Waging War on the American Dream, and How to Fight Back – Viking / 2006

D'Souza, Dinesh – What's So Great About America – Regnery Publishing / April 2002

Eccles, John C. - Evolution of the Brain: Creation of the Self - Routledge / 1989

Eddington, Arthur - Space, Time and Gravitation: An Outline of the General Relativity Theory - Cambridge University Press / February 1987

Ehrlich, Paul R. – Human Natures: Genes, Cultures, and the Human Prospect – Penguin Books / 2000-2002

Ehrlich, Robert - Nine Crazy Ideas: A Few Might Even Be True - Princeton University Press / 2001

Ehrlich, Robert – 8 Preposterous Propositions: From the Genetics of Homosexuality to the Benefits of Global Warming – Princeton Press / 2003

Ehrman, Bart D. – Jesus Interupted: Revealing the Hidden Contradictions in the Bible (and Why We Don't Know About Them) – Harper One / 2009

Ehrman, Bart D. – Misquoting Jesus: The Story Behind Who Changed the Bible and Why – HarperCollins / 2005

Ehrman, Bart D. – Peter, Paul, and Mary Magdalene: The Followers of Jesus in History and Legend – Oxford University Press / 2006

Ehrman, Bart D. – Truth and Fiction in the DaVinci Code: A Historian Reveals What We Really Know About Jesus, Mary Magdalene, and Constatine – Oxford University Press / 2004

Eldridge, Niles – Why We Do It: Rethinking Sex and the Selfish Gene – WW Norton & Company / 2004

Farmelo, Graham (Editor) – It Must Be Beautiful: Great Equations of Modern Science - Granta Books / March 2002

Feiler, Bruce – Where God was Born: A Journey by Land to the Roots of Religion – Harper Collins Publishing / 2005

Feiler, Bruce – Walking the Bible: A Journey by Land Through the 5 Books of Moses – Perennial / 2001

Feldman, Burton – 112 Mercer Street: Einstein, Russel, Godel, Pauli, and the End of Innocense – Arcade Publishing / 2007

Ferguson, Kitty - **Measuring the Universe** - Walker and Company / July 1999

Feynman, Richard P. - **Six Not-So-Easy Pieces:** Lectures on Symmetry, Relativity and Space-Time - Addison Wesley Longman, Inc. / February 1997

Feynman, Richard P. & Davies, Paul - **Six Easy Pieces:** Essentials of Physics Explained by its Most Brilliant Teacher - Addison Wesley Longman, Inc. / October 1998

Feynman, Richard P. & Sands, Matthew L. & Leighton, Robert B. - **The Feynman Lectures on Physics:** Commemorative Issue (3 Volume Set) - Addison Wesley Longman, Inc. / March 1994

Feynman, Richard P. - **QED:** The Strange Theory of Light and Matter - Princeton University Press / November 1988

Finkbeiner, Ann – **The Secret History of Sciences Past War Elite** – Viking Press / 2006

Flannery, Sarah – **A Woman's Mathematical Journey in Code** – Algonquin Books of Chapel Hill / 2002

Ford, Kenneth W. – **The Quantum World:** Quantum Physics for Everyone – Harvard University Press / 2004

Fox, Karen C. & Keck Aries – **Einstein A to Z** – John Wiley & Sons / 2004

Francis, Alan - **Everything Men Know About Women** - Andrews & McMeel / 1995

Friedman, Richard Elliot – **Who Wrote the Bible** – Harper San Francisco / 1997

Frymer-Kensky, Tikva – In The Wake of The Gods: Women, Culture, and the Biblical Transformation of Pagan Myth

Gabriel, Brigette – Because They Hate: A Survivor of Islamic Terror Warns America – St. Martins Press / 2006

Galison, Peter – Einstein's Clocks, Poincaré's Maps: Empires of Time – WW Norton & Company / 2004

Gamow, George - One Two Three...Infinity: Facts and Speculations of Science - A Mentor Book / October 1988

Gamow, George - The Great Physicists from Galileo to Einstein - Dover Publications, Incorporated / October 1988

Gardener, Lawrence – The Magdalene Legacy: The Jesus & Mary Bloodline Comspriracy – Barnes and Noble Books / 2005

Gardener, Martin - Relativity Simply Explained - Dover Publications, Incorporated / November 1996

Gardner, Martin - Weird Water & Fuzzy Logic: More Notes of a Fringe Watcher - Prometheus Books - Prometheus Books / October 1996

Gee, Henry – Jacobs Ladder: The History of the Human Genome – WW Norton & Company / 2004

Gertz, Bill – Break Down: How America's Intelligence Failure Led To September 11 – Regnery Publishing, Inc. / 2002

Gibbins, John - In Search of Schrödinger's Cat: Quantum Physics and Reality - Bantam Doubleday, Dell Publishing Group / September 1984

Goldberg, Bernard – Arrogance: Rescuing America From The Media Elite – Warner Books / 2003

Goldberg, Bernard - Bias: A CBS Insider Exposes How the Media Distort the News – Regnery Publications, Inc / 2002

Goldberg, Jonah – Liberal Fascism: The Secret History of the American Left From Mussolini to the Politics of Meaning – Doubleday / 2007

Goldsmith, Donald - The Runaway Universe: The Race to Find the Future of the Cosmos - Perseus Publishing / January 2000

Goldstein, Rebecca – Incompleteness: The Proof of the Paradox of Kurt Godel – WW Norton & Company / 2005

Greene, Brian – The Best American Science and Nature Writing – Houghton Mifflin Company / 2006

Greene, Brian - The Elegant Universe: Superstrings, Hidden Dimensions, and the Quest for the Ultimate Theory - Vintage Books / March 2000

Greene, Brian - The Fabric of the Cosmos: Space, Time, and the Texture of Reality – Vintage Books / 2005

Greenspan, Nancy Thorndike – The End of the Certain World: The Life and Science of Max Born: The Nobel Physicist Who Ignited the Quantum Revolution – Basic Books / 2005

Gren, Ingrid Wickel – The Gene Master: How a New Breed of Scientific Entrepreneurs Raced for the Biggest Prize – Henry Hold and Company / 2002

Gribbin, John - The Search for Super Strings & Symmetry, and the Theory of Everything - Little, Brown & Company / December 1998

Gribbin, John & Mary – **Annus Mirabilis 1905:** Albert Einstein and the Theory of Relativity – Chamberlain Bros. / 2005

Gross, Paul R & Levitt, Norman & Lewis Martin W., Edited by – **Flight From Science And Reason** - New York Academy Of Sciences, Distributed by The John Hopkins University Press / 1995

Guth, Alan H. - **The Inflationary Universe:** The Quest for a New Theory of Cosmic Origins - Addison Wesley Longman, Inc. / March 1998

Haisch, Bernard – **The God Theory:** Universes, Zero Point Fields, and What's Behind It All – Weiser Books / 2006

Halpern, Paul – **The Great Beyond:** Higher Dimensions, Parallel Universes, and the Extraordinary Search for a Theory of Everything – John Wiley & Sons / 2004

Hamer, Dean H. & Copeland, Peter - **Living With Our Genes:** Why They Matter More Than You Think - Anchor Books / March 1999

Hamer, Dean H. – **The God Gene:** How Faith is Hardwired into Genes – Doubleday / 2004

Hannegraaff, Hank & Maier, Paul – **The DaVinci Code: Fact or Fiction** - Tyndale House Publishers / 2004

Hannity, Sean – **Let Freedom Ring:** Winning The War of Liberty Over Liberalism – Harper-Collins Press / 2002

Harold, Franklin M. - **The Way of the Cell:** Molecules, Organisms, and the Order of Life - Oxford University Press / 2001

Harpaz, Amos - **Relativity Theory:** Concepts and Basic Principles - A K Peters Ltd / May 1993

Haskin, Susan – **Mary Magdalen:** Myth and Metaphor – Kenedy & Kenedy / 1993

Hawking, Stephen W. - **A Brief History of Time** - Bantam Books, Incorporated / September 1998

Hawking, Stephen W. & Penrose, Roger - **The Nature of Space and Time** - Princeton University Press / February 1996

Hawley, R. Scott & Mori, Catherine A. - **The Human Genome:** A User's Guide - Harcourt Academic Press / 1999

Hazen, Robert M. - **Why Aren't Black Holes Black?**: The Unanswered Questions at the Frontiers of Science - Anchor Books / 1997

Hazleton, Lesley – **Jezebel:** The Untold Story of the Bible's Harlet Queen – Doubleday / 2007

Hecht, Jeff – **Understanding Fiber Optics** – Prentice Hall / 2002

Henning, Genz & Heusch, Karin - **Nothingness:** The Science of Empty Space - Perseus Publishing / October 1998

Herbert, Nick - **Quantum Reality:** Beyond the New Physics - Bantam Doubleday Dell Publishing Group / September 1984

Hoffman, Banesh - **Relativity and its Roots** - Dover Publications, Incorporated / February 1999

Hoffman, Banesh - **The Strange Story of Quantum** - Dover Publications, Incorporated / June 1980

Hoffman, Paul – **The Man Who Loved Only Numbers:** The Story of Paul Erdös and The Search For Mathematical Truth – Hyperion / 1998

Hogan, James P. – Kicking the Sacred Cow: Questioning The Unquestionable and Thinking The Impermissible – Simon and Schuster / 2004

Holden, Alan - The Nature of Solids - Dover Publications, Incorporated / May 1992

Horgan, John - The End of Science: Facing the Limits of Knowledge in the Twilight of the Scientific Age / Broadway Books / May 1997

Hoyle, Fred & Burbidge, Geoffrey & Narliker, Joyant V. – A different Approach to Cosmology: From Static Universe Through The Big bang Towards Reality - Cambridge University Press / April 2000

Icke, Vincent - The Force of Symmetry - Cambridge University Press / January 1994

Ingraham, Laura – Shut Up and Sing – Regnery Publishing Inc. / 2003

Isaacson, Walter – Einstein: His Life and Universe – Simon Shuster / 2007

Jacobovici, Simcha & Pellegrino, Charles – The Jesus Family Tomb: The Discovery, The Investigation, and The Evidence That Could Change History – Harper SanFrancisco / 2007

Jeans, Sir James – Physics and Philosophy - Dover / 1981

Johnson, George - Strange Beauty: Murray Gell-Mann and the Revolution in 20th-Century Physics - Knopf Alfred A / September 1999

Johnson, George – Miss Leavitt's Stars: The Untold Story of the Woman Who Discovered How to Measure the Universe – Atlas Books / 2005

Johnson, Phillip E. - **Darwin on Trial** – Intervarsity Press / 1993

Johnson, Phillip E. - **Defeating Darwinsim by Opening Minds** – Intervarsity Press / September 1997

Jones, Steve - **The Language of Genes:** Solving the Mysteries of Our Genetic Past, Present and Future - Anchor Books, Doubleday / September 1992

Jones, Steve – **Y: The Descent of Men:** Revealing the Mysteries of Maleness – Houghton Mifflin Co. / May 2003

Kaku, Michio – **Einstein's Cosmos:** How Albert Einstein's Vision Transformed Our Understanding of Space and Time – WW Norton & Company / 2004

Kaku, Michio - **Hyperspace:** A Scientific Odyssey Through Parallel Universes, Time Warps, and the Tenth Dimension - Doubleday & Company, Incorporated / February 1995

Kaku, Michio - **Introduction to Superstring & M-Theory** - Springer Verlag New York, LLC / January 1998

Kaku, Michio – **Physics of the Impossible:** A Scientific Exploration into the World of Phasers, Force Fields, Teleportation, and Time Travel – Doubleday / 2008

Kaku, Michio - **Quantum Field Theory:** A Modern Introduction - Oxford University Press, Incorporated / March 1993

Kaku, Michio & Thompson, Jennifer - **Beyond Einstein:** The Cosmic Quest for the Theory of the Universe - Doubleday & Company, Incorporated / September 1995

Kane, Gordon L. - **Supersymmetry:** Unveiling the Ultimate Laws of Nature - Perseus Books Group / April 2000

Kane, Gordon L. – The Particle Garden: Our Universe as Understood by Particle Physics – Helix Books / 1995

Kasser, Rudolphe; Meyer, Marvin; Wurst, Gregor – The Gospel of Judas – National Geogrphic / 2006

Kauffman, Stuart - Investigations - Oxford University Press / 2000

Kauffman, Stuart (Different Authors) - **At Home in the Universe** - Chinmaya Pubns West / July 1999

Keller, Evelyn Fox - The Century of the Gene - Harvard University Press / October 2000

King, Karen L. – Gospel of Mary of Magdala, Jesus, and the First Woman Apostle – Pole Bridge Press / 2003

Kirshner, Robert P. - The Extravagant Universe: Exploding Stars, Dark Energy, and the Accelerating Cosmos - Princeton University Press / 2002

Knight, Christopher & Butler, Alan – Solomon's Power Brokers: The Secrets of Freemasonry, The Church, and The Illuminati. – Watkins Publishing / 2007

Kohn, Bob – Journalistic Fraud: How The New York Times Distorts The News and Why It Can No Longer be Trusted – WND Books / 2003

Kraemer, Joel L. – Maimonedes: The Life and World of One of Civilizations Greatest Minds – Doubleday / 2008

Krauss, Lawrence M., Baker, Deborah (Editor) - Atom: An Odyssey from the Big bang to Life on Earth ... and Beyond / Little, Brown & Company / April 2001

Krauss, Lawrence M – Quintessence: The Mystery of the Missing Mass - Basic Books / Jan 2001

Kresney, Herbert – The Lost Gospel: The Quest for the Gospel of Judas Iscariot – National Geographic / 2006

Kurtz, Paul – Science and Religion: Are They Compatable – Prometheus Books / 2003

Kurzweil, Ray – The Singularity is Near: When Human Transcend Biology – Viking Press / 2005

LaPierre, Wayne & Baker, James - Shooting Straight: Telling the Truth about Guns in America - Regnery Publishing, Inc., / September 2002

Larsen, Michael – How to Write a Book Proposal – Writers Digest Books / 1987

Lasica, J. D. – Darknet: Hollywood's War Against the Digital Generation – John Wiley & Sons / 2005

Lederman, Leon & Teresi, Dick – The God Particle: If the Universe is the Answer, What Is the Question? / Dell Publishing Company, Incorporated / December 1993

Leloup, Jean-Yues – The Gospel of Philip: Jesus, Mary Magdalene & The Gnosis of Secret Union – Inner Traditions / 2003

Lerner, Eric J. - The Big bang Never Happened: A Startling Refutation of the Dominant Theory of the Origin of the Universe – Vintage Book / 1992

Lessig, Lawrence – Free Culture: How Big Media Uses Technology and the Law to Lock Down Technology and Control Creativity – Penquin Press / 2004

Levitt, Steve D. & Dubner, Stephen J. – **Freakonomics:** A Rouge Economist Explores the Hidden Side of Everything – HarperCollins / 2005

Lewontin, Richard C. - **The Triple Helix:** Gene, Organism, and Environment - Harvard University Press / 2000

Lightfoot, Neil R. – **How We Got the Bible** – MJF Books / 2003

Lightman, Alan - **Great Ideas in Physics:** The Conservation of Energy, The Second Law of Thermodynamics, The Theory of Relativity, Quantum Mechanics - McGraw Hill / January 2000

Limbaugh, David – **Persecution:** How Liberals are Waging a War Against Christiananity – Regnery Press / 2003

Lincoln, Don – **Understanding the Universe from Quarks to the Cosmos** – World Scientific Publishing Co. / 2004

Lindley, David - **Where Does The Weirdness Go:** Why Quantum Mechanics is Strange, but Not as Strange as You Think - Perseus Books, L.L.C. / February 1997

Lindley, David – **Uncertainty:** Einstein, Heisenberg, Bohr, and The Struggle For The Soul of Science – Doubleday / 2007

Livio, Mario - **The Accelerating Universe:** Infinite Expansion, The Cosmological Constant, The Beauty of the Cosmos - John Wiley & Sons / March 2000

Livio, Mario – **The Equation That Couldn't Be Solved:** How Mathematical Genius Discovered the Language of Symmetry – Simon and Schuster/2005

Livio, Mario – **The Golden Ratio:** The Story of Phi, The Worlds Most Astonishing Number – Random House / 2002

Lloyd, Seth – **Programming the Universe:** A Quantum Computer Scientist Takes on the Cosmos – Bonzoi Books / 2006

Loomis, Greg – **Pegasus Secrets** – Leisure Books / 2005

Lott, John R - **More Guns Less Crime:** Understanding Crime and Gun Control Laws, - The University of Chicago Press / 2000

Lott, John R - **The Bias Against Guns:** Why Almost Everything You've Heard About Gun Control is Wrong Regnery Publishing / February 2003

Loewen, James W. – **Lies My Teacher Told Me:** Everything Your American History Book Got Wrong – A Touchtone Book / 2007

Lutzer, Erwin W. – **The DaVinci Deception** – Tyndale House Publishers / 2004

Mack, Barton L. – **The Lost Gospel:** The Book Q & Christian Origins Harper-Collins Publisher / 1993

Maddox, Brenda - **Rosalind Franklin:** The Dark Lady of DNA – HarperCollins / October 1, 2002

Maddox, John Royden - **What Remains to Be Discovered:** Mapping the Secrets of the Universe, the Origins of Life, and the Future of the Human Race - Free Press / October 1998

Margolis, Lynn & Sagan, Dorian – **Aquiring Genomes:** The Theory of the Origin of the Species – Basic Books / 2003

Margolis, Lynn & Sagan, Dorian - **Microcomos:** Four Billion Years of Microbial Evolution - University California Press / May 1997

Marmet, Paul – **Absurdities in Modern Physics:** A Solution, or, A Rational Interpretation of Modern Physics – Editions du Nordir / 1993

Marmet, Paul – **Einstein's Theory of Relativity Versus Classical Mechanics** – Newton Physics Books (Self Published) / Jan 1997

Marshall, I. N. & Zohar, Danah & Peet, David - **Who's Afraid of Schrödinger's Cat:** An A-to-Z Guide to All the New Science Ideas You Need to Keep Up with the New Thinking - William & Company / July 1998

Magueijo, Joao - **Faster than the Speed of Light:** The Story of a Scientific Speculation – October 2002

Manji, Irshad – **The Trouble with Islam:** A Moslem's Call for Reform – Random House / 2003

Mayr, Ernst – **One Long Argument:** Charles Darwin and The Genesis of Modern Thought – Harvard University Press /1991

McFadden, Johnjoe - **Quantum Evolution:** The New Science of Life - W.W. Norton & Company / 2000

McMillan, Pricilla J – **The Rain of J Robert Oppenheimer and the Birth of the Modern Arms Race** – Viking Penquin / 2005

Meyer, Marvin – **Gospel of Mary: The Secret Tradition of Mary Magdalene, the Companion of Jesus** – Harper San Francisco / 2004

Meyer, Marvin – **The Gnostic Discoveries:** The Impact of Nag Hammadi Library – Harper San Francisco / 2005

Mlodinow, Leonard – **Feynman's Rainbow:** A Search for Beauty in Physics and in Life - Warner Books, Incorporated / May 2003

Mlodinow, Leonard – **The Story of Geometry From Parallel Lines To Hyperspace** - Simon Schuster / April 2002

Moalen, Sharon – **The Surprising Connection Between Disease and Longevity** – Harper Perennial / 2007

Moffet, John W – **Reinventing Gravity:** A Physicist Goes Beyond Einstein – Smithsonian Books / 2008

Moore, David S. - **The Dependent Gene:** The Fallacy of Nature vs. Nurture - W H Freeman & Co / 2002

Moore, John A. – **From Genesis to Genetics:** The Case of Evolution and Creationism – University of California Press / 2003

Morange, Michel & Cobb, Matthew (Translator) – **The Misunderstood Gene** - Harvard University Press / January 2000

Moreland, J. P. – **The Creation Hypothesis:** Scientific Evidence for a Intelligent Design – Intervarsity Press / 1994

Morris, Dick – **Off With Their Heads:** Traitors Crooks & Obstructionists in American Politics, Media, & Business – Regan Books / 2003

Morris, Richard - **The Evolutionists:** The Struggle for Darwin's Soul - W. H. Freeman and Company / 2001

Morris, Richard - **The Universe, the Eleventh Dimension, and Everything:** What We Know and How We Know It - Four Walls Eight Windows / September 1999

Nadeau, Robert & Kafatos, Manes - **The Non Local Universe** - Oxford University Press, Incorporated / December 1999

Nasar, Silvia – **A Beautiful Mind**: The Life of Mathematical Genius and Nobel Laureate John Nash – Touchtone Books / 1998

Neffe, Jurgen – **Einstein**: A Biography – Farrar, Straus, and Giroux / 2007

Netz, Riviel & Noel, William – **Archimedes Codex:** How a Medieval Prayer Book is Revealing the True Genius of Antiquities Greatest Scientist – DaCapo Press / 2007

Newman, Sharon – **The Real History Behind The DaVinci Code** - Berkley Books / 2005

Norby, Jeremy – **The Cosmic Serpent:** DNA and the Origin of Knowledge – Putnam Books / 1998

Nordhaus, Ted & Shellenberger, Michael – Break Through: The Death of Environmentalism to the Politics of Possibility - Houghton Mifflin Company / 2007

Oerter, Robert – **The Theory of Almost Everything:** he Standard Model, The Unsung Triumph of Modern Physics – Pi Press / 2006

Ohanian, Hans C. – **Einstein's Mistakes:** The Human Feelings of Genius – WW Norton & Co. / 2008

Olson, Carl E. & Missel, Sandra – **The DaVinci Hoax:** Exposing the Errors in the DaVinci Code – Ignatious Press / 2004

Olson, Steve – **Count Down:** Six Kids Vie for Glory at the World's Toughest Competition – Houghton Miffin Company / 2004

Olson, Steve – **Mapping Human History:** Gene, Race, and Our Common Origin – A Mariner Book / 2002

Osterholm, Michael T. & Schwartz, John - **Living Terrors:** What American Needs to Know to Survive the Coming Bioterrorist Catastrophe - Delta Books / 2000

Overman, Dean L. - **A Case Against Accident and Self Organization** - Wolfhart Pannenberg; Paperback / 2001

Pagels, Elaine – **The Gnostic Gospels** – Vintage Books / 1979

Pagels, Elaine & King, Karen L. – **Reading Judas: The Gospel of Judas and The Shaping of Christianity** – Viking Press / 2007

Palumbi, Stephen R. - **The Evolution Explosion:** How Humans Cause Rapid Evolutionary Change - WW Norton & Company, Inc. / 2001

Parfitt, Tudor – **The Lost Ark of the Covenant:** Solving the 2500 Year Old Mystery of the Fabled Biblical Ark – HarperCollins / 2008

Parismo, Joseph – **Roosevelts Secret War:** FDR and WW2 Espionage – Random House / 2002

Park, Robert - **Voodoo Science:** The Road from Foolishness to Fraud - Oxford University Press, Incorporated / May 2000

Parker, Barry – **Albert Einstein's Vision:** Removable Discoveries That Stopped Modern Science – Prometheus Books / 2004

Parker, Barry – **Einstein:** The Passion of a Scientist – Prometheus Books / 2003

Peat, David – From Certainty to Uncertainty: The Story of Science and Ideas in The Twentieth Century – Joseph Henry Press / 2002

Peat, David - Superstrings and the Search for the Theory of Everything - NTC Publishing Group / July 1989

Peirce, B. O. - A Short Table of Integrals - Ginn and Company / 1929

Pelikan, Jaroslav – Whose Bible Is It?: The History of the Scriptures Through the Ages – Viking / 2005

Peters, F.E. – The Children of Abraham, Judaism, Christianity, Islam – Princeton University Press / 2004

Picknett, Lynn – Mary Magdalene: The True Life Story of the Legendary Figure Featured in the DaVinci Code – Graf Publishers / 2003

Picknett, Lynn & Prince, Olive – Templar Revelation: Secret Guardians of the True Identity of Christ – Toachtone / 1998

Pinker, Steven - How the Mind Works - W.W. Norton & Company / January 1999

Pinker, Steven – The Blank Slate: The Modern Denial of Human Nature – Penguin Books / 2002

O'Shea, Daniel – The Poincaré Conjecture: In Search of the Shape of the Universe – Walker & Company

Polkinghorne, John C. – Quantum World - Princeton University Press / 1984

Polkinghorne, John C. - Faith, Science and Understanding - Yale University Press / August 2001

Posner, Richard A. – Not a Suicide Pact: The Constitution in a Time of National Emergency – Oxford University Press Inc. / 2006

Poundstone, William – Prisoner's Dilemma: John von Neumann, Game Theory and the Puzzle of the Bomb

Powell S. Corey - God in the Equation: How Einstein Became the Prophet of the New Religious Era – Free Press / 2003

Randall, Lisa – Warped Passages: Unraveling the Mysteries of the Universe's Hidden Dimensions – HarperCollins Publishing / 2005

Rees, Martin – Just Six Numbers: The Deep Forces That Shape the Universe – Basic Books / 2002

Rees, Martin – Before The Beginning: Our Universe and Others – Perseus Books/August 1998

Rensberger, Boyce – Life Itself: Exploring the Realm of the Living Cells – Oxford University Press / 1996

Ridley, Matt – The Agile Gene: How Nature Turns on Nurture - HarperCollins / July 2004

Ridley, Matt - Genome: An Autograph of a Species in 23 chapters - HarperCollins - / October 2000

Ridley, Matt - The Red Queen: Sex and the Evolution of Human Nature - Penguin USA / June 1995

Ridley, Matt – Nature via Nurture: Genes, Experience, and What Makes Us Human – Harper Collins Publishing Inc. / 2003

Ridley, Matt – Origin of Virtue: Ridley, Matt – Human Instincts and the Evolution of Cooperation – Viking Penguin. / 1998

Rigden, John S. - Hydrogen: The Essential Element / Harvard University Press / May 2002

Rigden, John S. – Einstein 1905: The Standard of Greatness – Harvard University Press / 2005

Robinson, James M. – The Secrets of Judas: The Story of the Misunderstood Disciple and His Lost Gospel – Harper San Francisco / 2006

Rockmore, Dan – Stalking the Riemann Hypothesis: The Quest to the Hidden Law of Prime Numbers – Pantheon Books / 2005

Rose, Morris Edgar - Elementary Theory of Angular Momentum - Dover Publications, Incorporated / April 1995

Rosenberg, David – Abraham: The First Historical Biography – Basic Books / 2006

Rosenblum, Bruce - Quantum Enigma: Physics Encounters Consciousness – Oxford University Press / 2006

Rothman, Tony – Everything's Relative: And Other Fables from Science and Technology – John Wiles And Son / 2003

Rothman, Tony & Sudarshan, George & Comins, Shannon K. - Doubt and Certainty: The Celebrated Academy: Debates on Science, Mysticism, Reality, in General on the Knowable and Unknowable - Perseus Press / November 1999

Roughgarden, Joan – Evolution's Rainbow: Diversity, Gender, and Sexuality in Nature & People – California Press / 2004

Sagan, Carl – The Varieties of Scientific Experience: A Personal View of the Search For God – Penguin Press / 2006

Sapolsky, Robert M. – **The Trouble with Testosterone:** And Other Essays on the Biology of the Human Predicament – A Touchtone Book / 1997

Sartori, Leo - **Understanding Relativity** - Dover Publications, Incorporated / February 1991

Schaberg, Jane – **The Resurrection of Mary Magdalene:** Legends, Apocrypha, and the Christian Testament – Continuum / 2002

Schonfield, Hugh J. – **The Passover Plot** – Bantom Books / 1967

Schwartz, Jeffrey H. - **Sudden Origins:** Fossils, Genes, and the Emergence of Species - Hardcover / April 1999

Schwartz, Gary E.R. & Russek, Linda G – **Living Energy Universe** - Hampton Roads Publishing / 1999

Sears, Francis W. & Zemansky, Mark W. - **College Physics, Part1 and Part 2** - Addison Wesley Longman, Inc. / January 1987

Seife, Charles – **Alpha & Omega:** The Search for the Beginning and End of the Universe – Penguin Books / 2003

Seife, Charles – **Decoding The Universe:** How the New Science of Information Is Explaining Everything in the Cosmos, from Our Brains to Black Holes – Viking / 2006

Sepiro, George G. – **Poincaré's Prize:** The Hundred Year Quest to Solve One of Math's Greatest Puzzles – Dutton / 2007

Shlain, Leonard – **Sex, Time, and Power:** How Women's Sexuality Shaped Human Evolution – Penguin Books /2003

Shreeve, James – The Genome War: The Code of Life and How Craig Ventor Tried to Capture and Save the World _Alfred A. Knopf / 2004

Siegfried, Tom – Strange Matters: Undiscovered Ideas at the Frontiers of Spacetime – Joseph Henry Press / 2002

Silk, Joseph – The Big bang – W.H. Freeman & Company / 2001

Singer, Fred & Avery, Dennis T. – Unstoppable Global Warming: Every 1500 Years – Rowman and Littlefield Publisher Inc. / 2007

Smith, John Maynard & Szethmary, Eors - The Origin of Life: From the Birth of Life to the Origin of Language - Oxford University Press / May 1999

Smolin, Lee – The Trouble With Physics: The Rise of String Theory, The Fall of Science, and What Comes Next – Haughton Mifflin Company / 2006

Smolin, Lee - Three Roads to Quantum Gravity - Basic Books / February 2001

Spong, Shelby – Liberating the Gospel: Freeing Jesus from 2000 Years of Misunderstanding: Reaching the Gospel Through Jewish Eyes – HarperCollins / 1997

Spong, Shelby – Sins of the Scripture: Beyond Text of Hate to the God of Love – Harper San Francisco / 2005

Stachel, John – Einstein's Miraculous Year: Five Papers That Changed the Face of Physics – Princeton University Press / 1998

Stark, Rodney – Discovering God: The Origins of the Great Religions and Evolution of Belief: Mary Magdaline and the Holy Grail – Harper One / 2007

Starbird, Margaret – The Woman with the Alabaster Jar – Bear & Company / 2007

Stone, Merlin – When God Was a Women – Barnes & Noble Books / 1976 & 1993

Strobel, Lee – The Case for a Creator: A Jounalist Investigates Evidence That Points Toward a God – Zandevan / 2004

Susskind, Leonard - String Theory and the Illusion of Intelligent Design- Little Brown & Company / 2005

Susskind, Leonard – The Black Hole War: My Battle with Stephen Hawking to make the World Safe for Quantum Mechanics – Little, Brown, and Company / 2008

Sykes, Bryan – Adam's Curse: A Future Without Men - WW Norton and Company / 2004

Sykes, Bryan - The Seven Daughters of Eve: The Science That Reveals Our Genetic Ancestry - WW Norton and Company / 2001

Tabor, James D. – The Jesus Dynasty: The Hidden History of Jesus, His Royal Family, and the Birth of Chritiananity – Simon and Schuster / 2006

Tarcher, J P - The Cosmic Serpent: DNA and the Origin of Knowledge - Jeremy Nanby / April 1999

Tipler, Frank – The Physics of Immortality: Modern Cosmology, God and the Resurrection of the Dead - Doubleday & Company / August 1995

Teresi, Dick – Lost Discoveries: The Ancient Roots of Modern Science, from Babylonians to the Maya – Simon & Schuster / 2002

Torretti, Roberto – Relativity and Geometry - Dover Books on Mathematics / May 1996

Truss, Lynne – Eats, Shoots, and Leaves: The Zero tolerance Approach to Punctuation –Gotham Books / 2003

Tudge, Colin - The Impact of the Gene: From Mendel's Peas to Designer Babies - Farrar, Straus & Giroux, Incorporated / 2001

Tudge, Colin - The Time Before History: 5 Million Years of Human Impact – Touch Stone / 1996

Tyson, Neil DeGrrasse – Death by Black Hole: And Other Cosmic Quandaries – WW Norton & Company / 2007

Tyson, Neil DeGrasse & Goldsmith, Donald – Origin: Fourteen Billion Years of Cosmic Evolution – WW Norton Company / 2004

Unnikrishnan, C. S. - On the gravitational deflection of light and particles – Gravitation Group: Tata Institute of Fundamental Research (Mumbai, India) /Feb 2005 (On Internet)

Van Flandern, Tom – Speed of Gravity: What the Experiments Say -Meta Research, Univ. of Maryland Physics, Army Research Lab 6327 Western Ave., NW / Washington, DC 20015-2456 / 1998-02 (On Internet tomvf@metaresearch.org)

Vermes, Geza – Jesus the Jew: A Historian's Reading of the Gospel – Fortress Press / 1992

Vilar, Esther – **The Manipulated Man:** - Farrar, Straus and Giroux / 1972

Wade, Nichols (Edited by) - **The Science Times Book of Genetics** - The Lyons Press / January 1999

Walker, Mark & McKay, David - **Unraveling Genes**: A Lay Person's Guide to Genetic Engineering - Alen & Unwin / 2000

Wallace, Phillip R - **Paradox Lost**: Images of The Quantum - Springer / 1996

Wallace, Tim and Hopkins, Murphy & Marilyn – **Rosslyn:** Guardian of the Secrets of the Holy Grail – Barnes & Noble Books / 1999

Ward, Peter D & Brownlee, Donald - **Rare Earth**: Life is Uncommon in the Universe - Copernicus Books / January 2000

Weinberg, Steve - **Dreams of a Final Theory**: The Scientist's Search for the Ultimate Laws of Nature - Vintage Books / December 1993

Welborn, Amy – **Decoding DaVinci:** TheFacts Behind the Fiction of the DaVinci Code - OurSunday Visitors / 2004

Wells, Jonathan - **Icons of Evolution**: Science or Myth - Regnery Publishing, Inc. / 2000

Wells, Spencer – **Inside The Genographic Project:** The Landmark DNA Quest to Decipher Our Distant Pass – National Geogrphic Society / 2006

Wells, Spencer – **The Journey of Man:** A Genetic Odyssey – Random House Trade Paperbacks / 2002

Wick, David - The Infamous Boundary: Seven Decades of Heresy in Quantum Physics - Copernicus: Ann Imprint Of Springer-Verlag / 1996

Witherington III, Ben – The Gospel Code: Novel Claims About Jesus, Mary Magdalene and Da Vinci – Intervarsity Press / 2004

Woit, Peter – Not Even Wrong: The Failure of String Theory and Search for Unity in Physical Law – Basic Books / 2006

Wolf, Fred Alan - Parallel Universe: The Search for Other Worlds - Touchstone Books / 1988

Wolfram, Stephen – A New Kind of Science - Wolfram Media, Incorporated / May 2002

Wolfson, Richard – Simply Einstein: Relativity Demystified– WW Norton & Company / 2003

Wylin, Stephen M. – Jews in the Time of Jesus, An Introduction – Paulist Press / 1996

Yourgrau, Palle – A World Without Time: Godel & Einstein – Basic Books / 2005

Zahavi, Amatz & Avishey & Balaban, Amir - The Handicap Principle: A Missing Piece of Darwin's Puzzle - Oxford University Press / March 1999

Zebrowski, Ernest - A History of the Circle: Reasoning and the Physical Universe - Rutgers University Press / September 2000

Zee, Anthony - Fearful Symmetry: The Search for Beauty in Modern Physics (Princeton Science Library) - Princeton University Press / September 1999

Zimmerman, Barry E. & David – Why Nothing Can Travel Faster Than Light: And Other Explorations in Nature's Curiosity Show – Contemporary Books / 1993

Zohar, Danah - The Quantum Self: Human Nature and Consciousness Defined by New Physics

Zweiger, Gary - Transducing The Genome: Information, Anarchy, and Revolution in the Biomedical Sciences - McGraw Hill / 2001

INDEX

3-sphere, 47, 145, 213, 215, 216, 221, 227, 242, 250, 251
Abbott, Edwin, 42, 87, 88, 349
Accelerators, 81, 306
Action at a distance, 38, 62, 150, 171, 249, 301, 304, 342, 344
Aether, xv, 49, 149, 150, 241, 271, 272, 273, 274, 276, 277, 280, 293, 311
Allais, Maurice, xvi, 235, 236, 237
Angstrom, 84, 86, 93, 94, 177, 183, 212, 305
Angular momentum, 106, 194, 195, 196, 297
Angular velocity, 44, 48, 94, 106, 107, 108, 109, 111, 145
Anthropic Principle, xix, 314
Anti-matter, xix
Appearance, 38, 40, 42, 47, 56, 57, 86, 216, 221, 247

Aristarchus, 319
Aristotle, 50, 293, 317, 319
Aspect, Alain, 75, 249, 302
Atomic proportions, 42, 43, 75, 84, 93, 99, 213
Attraction, xviii, 44, 49, 52, 61, 67, 70, 71, 76, 97, 102, 104, 112, 127, 140, 170, 176, 177, 179, 180, 184, 186, 189, 192, 205, 208, 209, 211, 216, 217, 220, 226, 229, 230, 233, 236, 238, 266, 322, 325, 330, 333
Avogadro, Amedeo, 278, 295
Background radiation, xv, 77, 209, 210, 224, 250, 251, 331
Balloon, 78, 210, 216, 334
Bekenstein, Jacob, 231
Bell, John S., 51, 300, 302
Bending, xviii, 38, 43, 49, 64, 65, 71, 99, 100, 117, 121, 124, 125, 211, 232, 233, 245, 292, 307

Bending of light, xviii, 49, 71, 232, 233, 292
Beta decay, 308
Big bang, xiii, xv, 46, 47, 52, 76, 77, 78, 112, 179, 207, 208, 210, 211, 213, 215, 216, 219, 224, 310, 314, 324, 325, 330, 331, 334
Big crunch, 207
Big push, xiii, 47, 216, 235, 250
Biggest blunder in my life, 206, 323
Binary star, 140, 141, 241, 242
Bohr, Neils, 50, 51, 72, 190, 191, 193, 194, 197, 242, 249, 294, 295, 296, 297, 299, 301, 302, 304, 366
Boomerang, 70
Born, Max, 72, 150, 356, 359
Bose-Einstein Statistics, 189
Boson, 46, 188, 189, 308
Brown, Robert, 278, 351, 353, 354, 359, 364, 376
Brownian Motion, 278
Calculus, 52, 266, 322, 350
Caltech, 55, 228, 313
Camera obscura, 293
Cantor, George, 90
Causal contact, 77, 210, 221, 331
Causal Events, 283

Centrifugal, 48, 67, 109, 127, 164, 190, 205, 229, 231, 290
Centripetal, 67, 205
Cepheid variables, 327, 328
Challenger disaster, 55
Chandrasekhar, Subrahmanyan, 328
Clusters, 204, 211, 216, 232, 251, 323
Coiled, 94, 168
Collapses, 302
Colors, 294, 306, 329
Common sense, 38, 56, 62, 131, 284
Compactified, 74, 202
Conservation, 180, 217, 234, 315, 366
Conservation law, 315
Conservation of energy, 180, 217, 234, 315
Conservation of momentum, 315
Contracting, 76, 206, 209, 229, 323, 331
Coordinates, 98, 136, 137, 139, 143, 144, 150, 151, 152, 157, 168, 242, 268, 269, 280, 281, 282
Copenhagen interpretation, 72, 344
Copernicus, 52, 241, 317, 318, 319, 321, 378, 379
Corpuscular theory, 50, 294

Correlation, 249, 304, 332, 342
Cosmic Background Explorer, 209, 250
Cosmic microwave, 77, 209, 331
Cosmic microwave background, 77, 150, 209, 224, 250, 331
Cosmology, xiii, xix, 46, 47, 49, 52, 112, 145, 148, 204, 208, 225, 317, 319, 362, 377
Covariance, 49, 158, 179, 268, 271, 276, 277, 297, 311
Curled, 74, 314
Cylinder, 38, 43, 94, 96, 98, 99, 168
Dalton, John, 294
Dark matter, xv, xix, 48, 126, 205, 228, 229, 230, 232, 233, 234, 238, 335
De Broglie, Louis, 50, 191, 193, 298
Deferent, 318
Delayed choice experiment, xvii, 48, 53, 243, 246, 342
Deterministic, 54
Diffraction, xvi, 50, 71, 172, 175, 273, 294
Dirac, Paul, 55, 179, 180, 181, 182, 189, 193, 194, 296, 297, 299, 307, 332

Distance contraction, 44, 49, 63, 165, 280, 284
Diurnal, 318
Doppler effect, 47, 137, 141, 151, 156, 160, 204, 223, 229, 242, 250, 268, 269, 270, 281, 288, 329, 330
Duality, 41, 46, 50, 65, 66, 71, 72, 232, 294, 298, 299
Earth, xii, 49, 52, 53, 61, 63, 67, 68, 69, 76, 102, 104, 108, 112, 124, 141, 142, 143, 144, 146, 151, 156, 157, 161, 162, 163, 166, 169, 179, 204, 212, 215, 218, 219, 223, 224, 226, 230, 233, 238, 241, 242, 244, 250, 266, 267, 268, 272, 274, 281, 283, 284, 293, 307, 308, 317, 318, 319, 320, 321, 322, 324, 326, 327, 328, 330, 332, 335, 338, 339, 340, 341, 364, 378
Eddington, Arthur, 292, 355
Effective mass, 129, 148
Ehrenfest, Paul, xvii, 164
Einstein, xii, xvi, 38, 49, 50, 51, 52, 53, 60, 61, 62, 64, 66, 80, 92, 125, 126, 131, 149, 150, 165, 182, 189, 190, 191, 206, 225, 232, 242, 249, 260, 276, 277, 278, 279, 280, 285, 290, 291, 292, 294, 295, 298,

301, 304, 307, 311, 312, 315, 322, 323, 325, 338, 349, 351, 356, 357, 358, 359, 362, 363, 366, 367, 368, 369, 370, 371, 372, 373, 376, 379
Electromagnetic force, 45, 68, 70, 183, 185, 187, 189, 235, 308, 309, 311
Electron spin, 194, 195, 196, 297
Electron Spin, 194
Electrons, xvii, 45, 46, 50, 51, 62, 72, 97, 167, 170, 171, 172, 175, 179, 180, 183, 187, 189, 190, 191, 192, 193, 194, 195, 196, 209, 228, 277, 295, 296, 297, 298, 299, 300, 301, 302, 304, 305, 306, 309
Electro-optic delay, 343, 345
Electroweak, 308, 310, 311, 313
Elementary particles, 181, 306, 311, 314
Eleven-dimensional, 90
Energy, xii, xvi, xvii, xix, 38, 40, 42, 44, 46, 47, 50, 59, 61, 62, 78, 80, 82, 101, 104, 105, 106, 110, 111, 113, 117, 120, 129, 145, 146, 147, 148, 151, 157, 158, 159, 160, 163, 178, 179, 182, 188, 190, 191, 192, 193, 194, 196, 198, 212, 215, 217, 221, 224, 225, 234, 238, 239, 240, 248, 251, 253, 268, 270, 277, 279, 281, 287, 289, 295, 296, 297, 306, 315, 327, 332, 334, 335, 351, 364, 366, 374
Energy mass, 38, 42, 60
Energy-matter equivalence, 62, 80
Enigma, 38, 92, 304, 374
Entangled particles, 38, 41, 42, 45, 51, 53, 62, 75, 167, 169, 170, 171, 183, 187, 195, 244, 249, 250, 300, 301, 302, 304, 305, 342, 344, 347, 349, 352, 353
Epicycles, 317
EPR paradox, 51, 301, 302, 304
Equivalence of inertial, gravitational and energy mass, 42
Equivalence principle, 64, 120, 121, 126, 127, 231, 232, 246, 292
Ether, see aether, 271
Euclidean, 38, 42, 43, 45, 46, 56, 78, 80, 82, 86, 90, 91, 93, 94, 96, 99, 100, 101, 104, 105, 114, 117, 160, 168, 170, 171, 172, 187, 211, 215, 219, 221, 253, 260, 313, 332
Exchanging photons, 70, 71, 308
Expanding, xv, 46, 47, 52, 76, 79, 103, 206, 207, 210, 211, 213, 215, 216, 219,

220, 221, 225, 227, 229, 288, 323, 324, 333, 334, 349
Fermi particles, 46, 188
Fermi-Dirac statistics, 189, 307
Feynman diagram, 179, 180
Feynman, Richard, xi, xii, 55, 56, 102, 169, 173, 179, 180, 181, 244, 309, 357, 368
Fiber optic cable, 99, 302, 343
Figure !0.1, Parallax Shift, 327
Figure 10.2, Three Universe Expansion Solutions, 333
Figure 12.1, Instant Messaging, 342, 345, 347
Figure 12.1, Instant Messaging – Faster Than Light Speed Transmission Configuration, 345, 347
Figure 3.1, Four Dimensional Hyper-Box, 85
Figure 3.2, Flatlander Soldier Class, 88
Figure 3.3, Different Flatliner's Terrain, 89
Figure 4.1, Representation of One Axis, 95
Figure 4.2, Three-Dimensional Space Cells, 98
Figure 4.3, The Gyroscope Mechanism, 103, 107
Figure 4.4, Vectorial Representation Of The Fourth Axis, xxv, 103, 107, 115, 122
Figure 5.1, Simultaneity I - Special Relativity vs. Gyroverse Relativity, 133
Figure 5.1, Simultaneity I – System Relativity vs. Gyroverse Relativity, 133
Figure 5.2, Simultaneity II - The Gyroverse Vector Diagram, 135, 136
Figure 5.3, Simultaneity III – Speeding Bullet, 138, 139
Figure 5.4, Binary Star System, 141
Figure 5.4, Binary Star System, 141
Figure 5.5, Binary Star System Vector Diagram, 142, 143
Figure 5.5, Binary Star System Vector Diagram, 142, 143
Figure 5.6, Velocity Transformation, 154
Figure 5.6, Velocity Transformation, 154

Figure 6.1, Representation of One Axis, 168
Figure 6.2, Three-Dimensional Space, 170
Figure 6.2, Three-Dimensional Space Diagram, 170
Figure 6.3, Double Slit Experiment, 174
Figure 6.4, The Two Photon Paths, 177
Figure 6.5, Graviton Derivative Forces, 186
Figure 7.1, Three Universe Expansion Scenarios, 207
Figure 7.2, The Universe Model, 214
Figure 7.3, Virtual Location of Objects, 218, 227
Figure 7.4, Gravitational Lensing by Galaxy, 233
Figure 7.4, Virtual & Actual Path of Light, 226, 227
Figure 7.5, Cosmic Scale Delayed Choice Experiment, 237, 239, 245
Figure 7.5, Gravitational Lensing by Galaxy, 233
Figure 7.6, Cosmic Scale Delayed Choice Experiment, 237
Figure 7.6, Delayed Choice Experiment, xxvi, 247, 256, 261
Figure 7.6, Delayed Choice Experiment Using The Mach-Zehnder Interferometer, xxvi, 247, 256, 261
Figure 7.7, Delayed Choice Experiment, 239, 245
Figure 7.7, Delayed Choice Experiment Using The Mach-Zehnder Interferometer, 239, 245
Figure 8.1, The Michelson-Morley Experiment Test Setup, 273
Figure 8.2, The Fizeau Experiment Test Setup, 275
Figure 8.3, The Proverbial Special Relativity Train, 285
FitzGerald, George, 276
Fizeau experiment, 45, 155, 287
Fizeau, Armand-Hippolyte-Louis, 45, 49, 155, 156, 274, 275, 276, 287
Flatland, 42, 87, 349
Flatness of space, 52, 76, 77, 78, 207, 208, 210, 211, 221, 330, 334, 335
Flavors, 306
Flyby Anomoly, 238, 283

Four forces of nature, 51, 73, 306, 311
Four-dimensional, 39, 47, 84, 85, 86, 96, 114, 136, 160, 212, 213, 215, 238, 248, 259, 312
Fourth direction, 43, 44, 47, 93, 94, 113, 116
Frame, 44, 45, 62, 93, 98, 110, 114, 134, 136, 137, 139, 143, 144, 146, 148, 149, 150, 152, 156, 158, 160, 161, 165, 242, 272, 280, 282, 283, 284, 285, 287, 288, 290, 339, 340, 341
Friedmann, Alexander, 323, 325, 331
Full space, 97, 110, 111, 170, 249
Galaxy, 76, 127, 145, 204, 215, 216, 218, 221, 226, 227, 229, 230, 231, 232, 233, 234, 243, 244, 245, 323, 324, 328, 329, 335
Galilean relativity, 241, 267, 273
Galilean transformation, 49, 156, 160, 267, 268, 269, 271, 276, 279, 280, 281, 315
Galileo, Galilei, 52, 92, 266, 267, 320, 321, 358
Galois, Evariste, 316
Gamma factor, 44, 45, 118, 129, 146, 147, 150, 151, 154, 158, 159, 160, 163, 165, 166, 182, 226, 250, 253, 271, 281, 288, 298, 341
Gamow, George, 325, 358
Geiger counter, 300
Gell-Mann, Murray, 310, 362
General theory of relativity, xvii, xviii, 44, 46, 48, 49, 61, 62, 63, 64, 65, 66, 69, 79, 80, 92, 104, 121, 125, 126, 127, 165, 206, 225, 230, 231, 232, 233, 234, 235, 236, 238, 246, 259, 290, 291, 292, 307, 312, 322, 323, 325, 331, 335
Generation, 253, 306, 365
Geocentric, 52, 204, 317, 318
Geometric, 74, 88, 90, 113, 116, 316
Gerber, Paul, 291
Gerlach, Walter, 196
Gisin, Nicolas, 75, 302, 304, 342, 344
Glashow, Sheldon, 310, 311
Goudsmit, Samuel, 194, 297
Gravitational energy, 252, 332, 334
Gravitational force, 47, 48, 62, 63, 65, 67, 68, 69, 76, 96, 100, 101, 112, 121, 122, 123, 127, 183, 184, 185, 187, 205, 215, 224,

229, 230, 234, 236, 251, 290, 314, 334
Gravitational lensing, 232, 233
Gravitational mass, 61, 64, 92, 102, 105, 290
Graviton, xviii, 62, 65, 66, 67, 68, 100, 102, 103, 104, 111, 112, 183, 184, 186, 188, 216, 220, 222, 237, 238, 240, 253, 306, 307, 314
Gravity, xvii, xviii, 45, 46, 47, 48, 49, 51, 61, 63, 65, 66, 67, 68, 69, 70, 77, 80, 92, 94, 100, 101, 102, 103, 104, 105, 113, 120, 121, 123, 125, 127, 169, 175, 183, 185, 190, 206, 207, 211, 216, 217, 222, 224, 225, 230, 231, 234, 235, 237, 238, 239, 248, 251,259, 267, 290, 305, 307, 308, 309, 311, 312, 313, 314, 317, 321, 322, 324, 333, 334, 335, 376, 378
Groups, 315, 355
Guth, Alan, 77, 210, 251, 331, 332, 333, 334, 360
Gyrohelix, 43, 44, 89, 90, 96, 97, 101, 105, 108, 109, 110, 111, 114, 115, 117, 118, 129, 131, 144, 145, 169, 171, 182, 183, 196, 212, 216, 217, 222, 231, 242, 253

Gyroscope, 103, 105, 106, 107, 108, 116, 117
Gyro-vector, 114, 116, 118, 134, 137, 139, 142, 144, 154, 157, 162, 163
Gyroverse, ii, v, xi, xii, xiii, xxi, 37, 38, 39, 41, 42, 43, 44, 47, 48, 49, 53, 57, 61, 63, 64, 70, 72, 73, 74, 75, 78, 91, 92, 93, 104, 105, 114, 116, 117, 131, 132, 133, 134, 140, 149, 150, 151, 160, 165, 166, 175, 181, 189, 190, 202, 211, 213, 217, 224, 225, 226, 230, 231, 234, 235, 238, 242, 244, 246, 248, 250, 252, 253, 260, 291, 313, 341, 344, 399
Half-silvered mirror, 243, 246, 247, 248, 275, 346, 347
Heisenberg, Werner, 54, 193, 296, 366
Heliocentric, 52, 204, 241, 266, 318, 319, 320
Helix, 38, 42, 43, 91, 94, 95, 96, 97, 98, 99, 100, 101, 102, 104, 105, 111, 113, 144, 168, 170, 171, 176, 177, 184, 185, 186, 211, 213, 353, 363, 365
Hidden variable, 51, 53, 249, 250
Higgs, Peter, xviii, 310
Hilbert, David, 103, 315, 353
Hipparchus, 317

Horizon problem, 52, 76, 77, 78, 207, 208, 209, 210, 211, 330, 331, 334
Hoyle, Fred, 325, 362
Hubble age, 221, 224
Hubble constant, 76, 221, 224, 324, 329
Hubble, Edwin, 52, 76, 206, 221, 224, 323, 324, 329
Huygens, Christiaan, 50, 294, 298
Hydrogen atom, 188, 191, 297, 298, 299
Hyper-box, 84, 85, 86, 212, 248
Hyper-cylinder, 38, 43, 94, 98, 99, 168
Hyper-shell, 213, 215, 216, 218, 219, 220, 221
Hyper-sphere, 213
Hyper-torus, 94
Imaginary grid, 98, 170
Inertial force, 127, 230, 290
Inertial frame, 44, 45, 49, 62, 93, 98, 100, 109, 110, 114, 134, 136, 137, 139, 143, 144, 148, 149, 150, 152, 156, 157, 158, 160, 163, 242, 250, 267, 268, 269, 272, 276, 277, 279, 280, 282, 283, 284, 285, 287, 288, 290, 315, 339, 340, 341
Inertial mass, 44, 47, 48, 60, 61, 62, 64, 92, 105, 231, 290, 311

Inertial plane, 95, 136, 139, 283, 286
Infinity, xix, 83, 86, 90, 117, 123, 156, 222, 358
Inflation model, xii, xv, xix, 40, 46, 52, 74, 77, 78, 79, 210, 225, 251, 330, 332, 333, 334
Infrared radiation, 271, 298
Interference, 50, 72, 172, 243, 245, 246, 294, 298, 299, 343, 344
Interferometer, 246, 247, 256, 261, 272, 273, 275
Internet, 57, 378
invariance, 75, 111, 158, 267, 315
Invariance, 315
Isotopes, 241, 305
Isotropic, 150, 205, 250, 323
Jupiter, 320
Kaluza, Theodor, 39, 51, 80, 259, 260, 312, 313, 314
Kaluza-Klein, 39, 51, 80, 312, 313, 314
Kepler, Johannes, 52, 266, 320, 321, 322
kinetic, xv, xvii, 40, 44, 60, 113, 146, 157, 158, 159, 160, 213, 215, 217, 234, 271, 287, 289
Kinetic energy, xv, xvii, 40, 44, 60, 113, 120, 146, 157, 159, 160, 213, 215, 217, 234, 287, 289
Kirshner, Robert, 225, 335, 364

Klein, Oskar, 39, 51, 80, 259, 312, 313, 314
Kronkite, 66
Law of gravitation, 52, 322
Law of inertia, 92, 266
Laws of nature, 113, 181, 249, 267, 279, 280, 315
Leavitt, Henrietta, 327, 362
Leibniz, Gottfried Wilhelm, 266, 322, 350
Leptons, 306, 307
LeSage theory, 69, 103, 105, 112, 216, 222, 236
LeSage, George-Louis, 69, 102, 103, 104, 105, 112, 216, 222, 236
Lie groups, 316
Light beam, 155, 273
Light year, 38, 42, 47, 53, 93, 97, 161, 204, 210, 221, 233, 243, 244, 284, 324, 328, 338, 340, 341
Lightspeed, xix, 40, 44, 53, 69, 108, 117, 137, 144, 156, 219, 276, 280, 304, 345
Linde, Andrei, 77, 210, 332
Lomonosov, Mikhail, 102
Lorentz transformation, 44, 45, 63, 145, 149, 150, 179, 241, 242, 276, 279, 280, 297, 311, 315
Lorentz, Hendrik, 44, 45, 63, 149, 150, 151, 164, 179, 241, 242, 276, 277, 279, 280, 281, 297, 311, 315

Luminosity, 52, 225, 326, 327, 328, 335
Mach, Ernst, 61, 92, 246, 247, 256, 261, 289, 290
MACHOS, 229
Mack, Ernst, 61, 92, 246, 247, 256, 261, 289, 290
Mack-Zehnder Interferometer, 246, 247, 256, 261
Magellanic cloud, 327
Magueijo, João, xii, 368
Manifold, 39, 42, 43, 90, 91, 93, 314
Marmet, Paul, 291, 367
Mass, xvii, xviii, 38, 42, 44, 45, 47, 48, 49, 59, 61, 62, 64, 75, 80, 82, 92, 105, 111, 112, 113, 121, 123, 125, 129, 140, 141, 145, 146, 148, 149, 151, 153, 157, 158, 159, 163, 187, 188, 205, 208, 212, 213, 216, 228, 231, 232, 233, 234, 236, 239, 240, 248, 251, 252, 253, 267, 268, 270, 276, 277, 278, 279, 281, 287, 289, 290, 305, 306, 307, 310, 328, 335, 364
Mass-energy equivalence, 158, 287, 289
Matter, xv, xvii, xix, 38, 43, 44, 47, 48, 50, 51, 54, 57, 59, 60, 61, 63, 66, 68, 74, 76, 77, 78, 79, 80, 91, 92, 94, 96, 97, 99, 100, 102,

105, 109, 110, 112, 113, 114, 116, 117, 121, 126, 127, 129, 144, 145, 146, 150, 157, 158, 165, 175, 177, 178, 180, 182, 183, 184, 188, 189, 191, 192, 193, 202, 204, 208, 209, 211, 212, 213, 215, 216, 217, 218, 219, 220, 221, 222, 223, 224, 225, 228, 229, 230, 231, 232, 233, 234, 235, 236, 239, 248, 250, 251, 252, 287, 288, 290, 291, 293, 294, 296, 298, 300, 304, 305, 307, 311, 313, 323, 325, 330, 331, 332, 333, 334, 335, 354, 357, 360

Maxwell, James Clerk, 49, 50, 271, 276, 279, 294, 298, 308, 311, 312

Mechanical advantage, xii, 101

Mediating force particle, xviii, 62, 65, 66, 67, 71, 102, 307, 308

Metaphor, 66, 67, 78, 82, 157, 171, 210, 212, 236, 360

Metaphysical, 55

Michelson, Albert Abraham, 49, 267, 272, 273, 274, 276, 280

Michelson-Morley, 49, 267, 272, 273, 274, 276, 280

Microwave, 77, 224, 250

Milgrom, Mordehai (Moti), 126, 230

Minkowski, Herman, 103

Minovich, Michael, 238

Modified gravity theory, 231

Moffat, Paul, 231

Momentum, 82, 106, 107, 109, 157, 180, 191, 193, 194, 195, 196, 266, 287, 297, 301, 315

MOND, 48, 126, 127, 230, 231, 232, 234, 235, 251

Monochromatic light, 151, 268, 277, 281, 295, 342

Moon, xvi, 48, 54, 69, 111, 204, 236, 237, 266, 307, 317, 320, 322

Morley, Edward, 49, 267, 272, 273, 274, 276, 280

Motion-direction, 47, 93, 94, 95, 96, 98, 100, 105, 109, 110, 113, 114, 115, 116, 117, 118, 131, 137, 139, 143, 145, 146, 148, 152, 154, 160, 161, 163, 170, 182, 196, 212, 214, 215, 216, 217, 218, 220, 222, 242

M-theory, 90, 314

Multiverse, 252, 314

Muon, 165, 166, 306

NASA, 235

Natural occurrences, 38, 140

Nebulae clusters, 204

Neutral atoms, 47, 209, 213
Neutrino, 306
neutrinos, 183, 228
Neutron star, 140, 141
Neutrons, 46, 51, 183, 189, 228, 240, 305, 306, 308, 310
New York City, 56
Newton, Isaac, xvii, 38, 49, 50, 52, 54, 61, 62, 68, 69, 92, 158, 194, 232, 266, 267, 268, 290, 294, 298, 307, 321, 322, 323, 350, 367
Nicolas Gisin, 75, 302, 304, 342, 344
Nine-dimensional, 98
Noether, Emma, 315
Non-commuting observables, 301
Non-simultaneity, xvii, 40, 44, 53, 63, 100, 131, 136, 140, 153, 277, 282, 283, 284, 287
Nuclear, 51, 146, 188, 190, 225, 229, 295, 308, 309, 310, 313, 325, 328
Nuclear force, 308, 310, 313
Nucleons, 172, 184, 185, 186, 187, 240, 305, 308
Occam, William of, 58, 59, 71, 180
Omega, 208, 330, 350, 375
Orthogonal dimensions, 91, 93, 94, 96, 98, 99, 112, 113, 117, 118, 121, 147, 152, 154, 224, 269, 273
Pack, 76, 172, 307
Packets, 172, 253
Paradigm, xi, xiii, xv, xxi, 66, 67, 69, 70, 78, 102, 104, 105, 114, 116, 126, 131, 195, 206, 217, 223, 224, 227, 230, 231, 276, 323
Paradox, xvii, 51, 52, 53, 72, 131, 134, 140, 164, 165, 225, 301, 304, 338, 359, 378
Parallax, 52, 204, 317, 326, 327, 328
Part I, The Gyroverse Model Theory, 37, 39, 40, 41, 265, 337
Part II, Current Physics Overview, 39, 40, 41, 265, 337
Part III, Thought Experiments, 39, 337
Pauli, Wolfgang, 192, 307, 356
Perihelion advance, xviii, 49, 69, 194, 291
Perlmutter, Saul, 225, 334
Phenomena, xi, xvii, xviii, 38, 42, 45, 50, 51, 53, 55, 60, 61, 62, 64, 71, 92, 109, 116, 121, 124, 140, 147, 149, 152, 153, 166, 171, 172, 178, 179, 189, 190, 210, 227, 231, 232, 234, 235, 238, 242, 244, 248, 249, 251, 277, 294, 297,

298, 301, 302, 304, 309, 311, 320, 342, 344, 352

Philosopher, 50, 58, 289, 293, 317

Photoelectric effect, 50, 277, 294, 295, 298

Photons, 46, 48, 51, 64, 70, 71, 72, 75, 80, 100, 112, 121, 123, 124, 131, 133, 134, 136, 137, 139, 148, 155, 167, 171, 172, 174, 175, 176, 177, 178, 179, 180, 181, 183, 189, 193, 194, 209, 215, 218, 219, 241, 243, 244, 245, 248, 253, 296, 300, 302, 303, 304, 308, 309, 342, 343, 344, 345, 346, 348

Physicists, xii, xix, 54, 55, 57, 67, 70, 74, 80, 90, 131, 150, 196, 267, 271, 278, 294, 308, 314, 315, 316, 332, 358

Physics, xi, xii, xv, xviii, xix, xxi, xxiii, 39, 40, 41, 42, 43, 44, 45, 47, 48, 49, 51, 54, 55, 56, 57, 58, 60, 65, 66, 71, 72, 73, 74, 77, 78, 79, 80, 81, 90, 126, 128, 149, 158, 164, 172, 173, 181, 190, 192, 193, 195, 196, 205, 210, 212, 217, 225, 230, 231, 234, 238, 248, 249, 253, 265, 266, 277, 278, 279, 289, 290, 295, 296, 298, 301, 309, 310, 311, 313, 314, 316, 321, 331, 335, 349, 353, 357, 358, 361, 362, 363, 366, 367, 368, 370, 374, 375, 376, 377, 378, 379, 380

Pioneer 10 & 11 spaceshots, xvi, 235

Pitch of Gyroverse, 44, 94, 111, 113, 144, 145, 217, 220, 222, 223, 270

Planck, Max, 50, 57, 75, 190, 191, 194, 294, 295, 297, 312

Planck-length, 75

Plane of rotation, 105, 108, 109, 110, 113, 114, 144, 229, 231

Planets, 52, 68, 190, 191, 194, 229, 238, 266, 291, 307, 317, 318, 319, 321

Podolsky, Boris, 301

Poincaré, Henri, 277, 279, 358, 372, 375

Polarization, 75, 300, 302, 303

Position axes, 215

Position-space, 93, 94, 96, 98, 100, 113, 117, 118, 131, 134, 139, 144, 145, 146, 148, 152, 154, 156, 157, 160, 161, 162, 170, 177, 197, 211, 215, 216, 218, 219, 220, 222, 242

Postdiction, 80, 81

Postulates of relativity, 279

Precess, 108, 109, 110, 117, 154, 217, 222, 231

Principia, 49, 52, 92, 266, 321
Probability function, 72, 299
Professor, xii, 55, 72, 126, 225, 228, 230, 241, 298, 311, 314, 315, 320, 332, 335
Prograde, 318
Proton, 45, 46, 51, 157, 179, 183, 187, 188, 189, 190, 196, 228, 240, 305, 306, 308, 310
Protons, 45, 46, 51, 179, 183, 187, 189, 228, 305, 306, 308, 310
Ptolemaic system, 318
Ptolemy, Claudius, 317, 319
QCD, quantum chromodynamics, 51, 310
QED, Quantum electrodynamics, 51, 55, 56, 309, 310, 357
Quantum entanglement, 38, 41, 42, 51, 53, 62, 167, 195, 300, 301, 302, 347
Quantum fluctuation, 210, 332
Quantum number, 191, 192, 193, 194, 195, 296
Quantum physics, xiii, xvii, 40, 41, 45, 49, 50, 51, 54, 55, 73, 74, 75, 77, 78, 79, 128, 131, 150, 164, 167, 172, 173, 179, 182, 190, 193, 210, 231, 242, 244, 249, 250, 278, 294, 295, 296, 299, 300, 301, 302, 303, 306, 309, 313, 332
Quarks, 306, 310
Quasars, 232, 233, 243, 245
Radar waves, 271, 298
Radio waves, 271, 298
Ratio, xii, 56, 59, 118, 145, 146, 150, 152, 163, 208, 212, 222, 223, 224, 246, 253, 289, 330, 366
Real universe, 40, 56, 79
Reality, 38, 40, 55, 75, 86, 95, 116, 172, 220, 299, 334, 354, 358, 359, 361, 362, 374
Red shift, also see Doppler effect, 47, 52, 141, 156, 206, 223, 226, 233, 326, 329, 330, 335
Reflection, 293, 294
Refraction, 156, 293, 294, 343
Relativity, xii, xiii, xv, xvi, xvii, xviii, xix, 40, 41, 43, 44, 45, 46, 48, 49, 53, 54, 60, 61, 62, 63, 64, 65, 66, 69, 73, 74, 76, 78, 79, 81, 92, 116, 118, 125, 127, 131, 132, 133, 134, 140, 149, 150, 153, 155, 156, 158, 160, 162, 164, 165, 166, 179, 182, 194, 195, 206, 210, 211, 225, 230, 231, 232, 233, 234, 235, 236, 241, 242, 246, 259, 266, 267, 273, 274, 276, 277, 278, 279, 280, 283,

284, 285, 286, 290, 291, 295, 297, 304, 307, 309, 311, 312, 322, 323, 325, 330, 331, 332, 334, 335, 338, 340, 341, 349, 351, 355, 357, 358, 359, 360, 361, 366, 367, 374, 377, 379

Repel, 70, 175, 178, 192

Repulsion mechanism, 47, 70, 71, 179, 189

Rest energy, 44, 60, 158, 196, 215

Rest frame, 45, 93, 146, 152, 161, 280, 282, 283

Retrograde, 318

Revolutions, 168, 170, 217

Roll, 44, 67, 94, 111, 118, 144, 145, 148, 213, 214, 220, 222

Roll/pitch ratio, 118, 145, 146, 148, 150, 152, 163, 212, 222

Roll-pitch product, 118, 145, 212, 213, 214, 217, 222

Rosen, Nathan, 301

Rubin, Vera Cooper, 229

Rutherford, Ernest, 50, 190, 193, 295

Salam, Abdus, 310, 311

Schrödinger, Erwin, 50, 73, 131, 150, 193, 194, 195, 296, 297, 298, 299, 300, 301, 304, 351, 358, 367

Schwinger, Julian, 56, 309

Sister, 75, 97, 98, 167, 168, 170, 171, 172, 175, 184, 186, 187, 244, 248, 300, 301, 344

Sister location, 97, 98, 168, 171, 172, 175, 184, 186, 187, 244, 248, 344

Slits, xvi, 46, 72, 73, 172, 175, 243, 245, 342

Smith and Dale, 66, 67

Smolin, Lee, 314, 375, 376

Solar, xvi, 52, 61, 69, 80, 190, 229, 236, 238, 266, 292, 307, 318, 322

Solar eclipse, 69, 80, 236, 292

Space-platform, 150, 151, 152, 153, 154, 157, 160, 268, 269, 270, 281, 282, 283, 284, 287, 288, 340

Space-time, 65, 66, 74, 90, 126, 140, 202, 259, 307, 312, 323

Special theory of relativity, xviii, 41, 43, 44, 45, 48, 49, 53, 59, 61, 62, 66, 69, 79, 116, 118, 131, 132, 134, 140, 149, 150, 153, 155, 156, 158, 160, 162, 164, 165, 166, 179, 182, 241, 242, 274, 276, 277, 279, 280, 283, 284, 286, 290, 291, 295, 297, 304, 312, 322, 340, 341

Speed of light, xii, xiii, xv, xviii, 38, 43, 46, 49, 53, 59, 60, 63, 64, 65, 69, 75, 79,

80, 91, 94, 95, 96, 100, 101, 104, 105, 108, 109, 113, 114, 115, 116, 118, 131, 132, 134, 136, 137, 139, 140, 141, 143, 144, 145, 148, 149, 151, 152, 155, 156, 159, 160, 161, 162, 164, 166, 170, 182, 183, 194, 195, 197, 202, 212, 214, 215, 219, 221, 222, 226, 241, 242, 251, 253, 267, 271, 272, 273, 274, 276, 277, 279, 280, 281, 283, 287, 288, 297, 300, 302, 303, 304, 307, 309, 329, 330, 338, 339, 341, 342

Spin, xiii, xix, 41, 46, 68, 179, 190, 194, 195, 196, 297, 300, 301, 302, 304, 307, 308

Spooky action at a distance, xvi, 38, 62, 249, 301, 304

Standard candle, 52, 204, 326, 327, 328

Standard model, 51, 305, 310, 311

Stars, xviii, 48, 52, 76, 127, 140, 146, 204, 208, 211, 215, 216, 219, 223, 225, 227, 229, 230, 233, 234, 241, 242, 317, 318, 321, 324, 325, 326, 327, 330, 331, 364

Static universe, 68, 70, 206, 238, 241, 253, 323, 335, 362

Steady state model, 52

Steinhardt, Paul, 77, 210, 332

Stellar parallax, 317, 326

Stern, Otto, 196

String theory, 39, 43, 51, 73, 74, 80, 81, 90, 91, 202, 203, 307, 313, 314

Strong force, 45, 51, 183, 184, 185, 186, 187, 305, 307, 308

Structure, ii, v, xiii, 38, 41, 42, 46, 51, 58, 72, 80, 93, 117, 189, 190, 191, 192, 193, 194, 195, 224, 226, 244, 249, 253, 299, 307, 319

Sun, xvi, xix, 48, 49, 52, 54, 64, 65, 68, 69, 80, 104, 122, 123, 124, 125, 126, 169, 178, 188, 190, 191, 194, 204, 216, 232, 235, 236, 237, 241, 250, 266, 272, 291, 292, 296, 307, 317, 318, 319, 320, 321, 322, 327, 328

Supernova, Type 1a, 225, 328, 335

Supersymmetry, 363

Susskind, Leonard, 314, 376

Symmetric, 141, 288, 315

Tau, 306

The Starry Messenger, 320

Three-dimensional, 38, 41, 42, 43, 46, 66, 67, 78, 80, 82, 83, 84, 86, 87, 89, 91,

93, 95, 96, 97, 98, 104, 110, 111, 113, 117, 150, 160, 168, 170, 171, 172, 183, 187, 202, 212, 213, 215, 220, 238, 244, 248, 249, 267, 344
Time dilation, 44, 45, 49, 63, 145, 149, 152, 276, 282, 284, 339
Time dimension, 43, 74, 90, 202, 259, 312
Timepiece, 129, 146, 153, 269, 285, 286, 340, 341
TOE, Theory of everything, 51, 313
Torus, 94, 211
Transitivity, 62, 284, 286
Transverse, 100, 217, 234
Treatise, xiii, 38, 47, 56, 58, 60, 68, 71, 80, 93, 192, 202, 249, 253, 266, 321
Triplet paradox, 52
Twelve-dimensional, 38, 42, 43, 45, 46, 93, 95, 97, 110, 114, 145, 150, 159, 168, 170, 171, 183, 184, 185, 187, 212, 240, 244, 252, 253, 344
Twin, xvi, 52, 338
Twin paradox, 52
Two-dimensional, 42, 66, 83, 85, 87, 88, 113, 165, 210, 248, 318, 332, 334
Uhlenbeck, George, 194, 297
Ultraviolet rays, 271, 298

Uncertainty principle, 77, 180
Understand before judging, 58
Unify physics, 51, 65, 73, 90, 193, 202, 259, 296, 307, 311, 312, 313
Universe, ii, v, xii, xiii, xv, xix, 38, 39, 40, 41, 42, 43, 46, 47, 48, 51, 52, 54, 55, 56, 57, 58, 61, 74, 75, 76, 77, 78, 79, 80, 82, 84, 85, 86, 90, 91, 92, 93, 94, 97, 98, 99, 102, 104, 109, 111, 112, 117, 127, 140, 141, 144, 145, 148, 167, 169, 170, 179, 181, 183, 203, 204, 206, 207, 208, 209, 210, 211, 212, 213, 214, 215, 216, 217, 218, 219, 220, 221, 222, 223, 224, 225, 226, 227, 228, 230, 231, 234, 235, 239, 241, 248, 249, 250, 251, 252, 253, 267, 272, 286, 290, 301, 307, 310, 311, 313, 314, 315, 317, 319, 322, 323, 324, 325, 328, 330, 331, 332, 333, 334, 335, 349, 350, 353, 354, 356, 359, 360, 362, 363, 364, 365, 366, 367, 369, 372, 373, 374, 375, 378, 379, 380, 399
University, 55, 72, 75, 102, 164, 193, 194, 225, 229, 246, 295, 298, 302, 310,

311, 314, 319, 320, 332, 335, 349, 350, 351, 352, 354, 355, 356, 357, 360, 361, 362, 363, 364, 365, 366, 367, 368, 369, 371, 372, 373, 374, 375, 376, 379, 380, 399
Unnikrishnan, C. S., 121, 378
Van de Graaff generator, 70
Vector, 114, 116, 118, 123, 131, 134, 135, 136, 137, 139, 142, 143, 144, 154, 162, 163, 234, 270, 306, 321
Velocity addition formula, 287
Velocity transformation, 45, 153, 276, 287
Venus, 320
Virtual particle, 180, 182
Wave, 50, 65, 71, 72, 73, 121, 124, 125, 126, 150, 160, 171, 172, 191, 193, 194, 196, 202, 232, 253, 271, 278, 288, 294, 295, 296, 297, 298, 299, 300, 302, 304, 312, 343
Wave function, 50, 72, 73, 150, 194, 297, 299, 300, 302, 304, 312

Wave packet, 171, 172, 196, 253
Weak force, 45, 51, 183, 186, 308, 309, 311, 312
Weinberg, Steven, 310, 311, 378
Wheeler, John, xvii, 48, 242, 243, 245, 246, 342, 344
Wheeler, John Archibold, xvii, 48, 242, 243, 245, 246, 342, 344
White dwarfs, 225, 328
Wilkinson microwave anistropy probe, 209, 224, 250
WIMPS, 228
Windings, 83, 95, 97
Woit, Peter, 314, 379
Wrapped, 38, 43, 98, 212, 252
Wright, Frank Lloyd, 56
x, y, and z, 43, 91, 93, 94, 96, 98, 99, 115, 131, 182, 212
Yaw, 44, 47, 144, 145, 146, 152, 154, 162, 163, 212, 216, 220, 222, 242, 341
Young, Thomas, 50, 172, 294, 298
Zwicky, Fritz, 228

The Gyroverse
The Hidden Structure of the Universe

Donald Wortzman

ABOUT THE AUTHOR

In addition to this third edition of The Gyroverse, Donald Wortzman, a Rotarian, recently wrote Daughters of Ishtar, a fiction thriller about a contemporary conspiracy, with roots going back over 4500 years to a matriarchal tribe, which aims to rework all mankind.

He also produced several videos on varied topics that can be viewed on YouTube, including two interviews for TV discussing his books and authoring in general.

While previously employed at IBM, he received many patents for his work. He also presented numerous original research papers, on a variety of technical subjects at major scientific conferences. These were conferences sponsored by The New York Academy of Sciences, Annual Conference on Engineering in Medicine and Biology, Numerical Control Society, and Eastern Joint Computer Conference.

He received his Bachelor's in Electrical Engineering from City College of New York and a Master of Science in Electrical Engineering from Syracuse University. He has also completed his course work toward a Ph.D. in Mathematics at New York University.

Most of his working life was in the field of engineering and system design. While it might seem odd that an engineer rather than a physicist would propose such a theory, this model of the universe and the rules that govern it are more of an engineering design than a mathematical formalism.